Behavioral Flexibility in Primates: Causes and Consequences

DEVELOPMENTS IN PRIMATOLOGY: PROGRESS AND PROSPECTS

Series Editor:

Russell H. Tuttle
University of Chicago, Chicago, Illinois

This peer-reviewed book series will meld the facts of organic diversity with the continuity of the evolutionary process. The volumes in this series will exemplify the diversity of theoretical perspectives and methodological approaches currently employed by primatologists and physical anthropologists. Specific coverage includes: primate behavior in natural habitats and captive settings; primate ecology and conservation; functional morphology and developmental biology of primates; primate systematics; genetic and phenotypic differences among living primates; and paleoprimatology.

ALL APES GREAT AND SMALL
VOLUME I: AFRICAN APES
Edited by Biruté M.F. Galdikas, Nancy Erickson Briggs, Lori K. Sheeran, Gary L. Shapiro and Jane Goodall

THE GUENONS: DIVERSITY AND ADAPTATION IN AFRICAN MONKEYS
Edited by Mary E. Glenn and Marina Cords

ANIMAL MINDS, HUMAN BODIES
By W.A. Hillix and Duane Rumbaugh

COMPARATIVE VERTEBRATE COGNITION
Edited by Lesley J. Rogers and Gisela Kaplan

ANTHROPOID ORIGINS: NEW VISIONS
Edited by Callum F. Ross and Richard F. Kay

MODERN MORPHOMETRICS IN PHYSICAL ANTHROPOLOGY
Edited by Dennis E. Slice

BEHAVIORAL FLEXIBILITY IN PRIMATES: CAUSES AND CONSEQUENCES
By Clara B. Jones

Behavioral Flexibility in Primates: Causes and Consequences

CLARA B. JONES

Fayetteville State University
Fayetteville, North Carolina

Theoretical Primatology Project
Fayetteville, North Carolina

Community Conservation, Inc.
Gays Mills, Wisconsin

 Springer

Library of Congress Cataloging-in-Publication Data

Jones, Clara B.
    Behavioral flexibility in primates : causes and consequences / Clara B. Jones.
        p.   cm.—(Developments in primatology)
    Includes bibliographical references (p.   ).
    ISBN 0-387-23297-4
    1. Primates—Behavior.   I. Title.   II. Series.

    QL737.P9J65 2005
    599.8'15—dc22   2004065094

ISBN 0-387-23297-4

Printed in the United States of America.

9 8 7 6 5 4 3 2 1

springeronline.com

Contents

Foreword

Some 50 years ago, researchers started a study on the behavior of Japanese macaques on the islet of Koshima near Japan (Kawai, 1965). To give the monkeys an incentive to emerge from the forest onto the beach they provided sweet potatoes and occasionally some wheat. In 1953, a young female called Imo started washing these sweet potatoes in water before eating them. This novel behavior was soon adopted by other members of the troop and spread through the population. When Imo was four years old, she discovered that by throwing a mix of wheat and sand in the sea she could separate the grains from the unwanted sand. Again, the behavior was imitated by other group members and, after a few years, most monkeys practiced this method of obtaining grains.

This well-known example of innovative behavior and its cultural transmission was one of the first to document primate behavioral flexibility in the field. It is not only in their foraging behavior that monkeys and apes display the most complex arrays of behaviors. For instance, Goodall's work in Gombe (e.g., van Lawick-Goodall, 1968) revealed many examples of behavioral patterns that never failed to surprise field researchers and the interested public alike. Only our own species surpasses other primates in exhibiting such a large repertoire of flexible responses in heterogeneous environments, a factor that certainly contributed to our ability to occupy almost any habitat. This striking similarity in the ability to show flexible behaviors makes primates, including humans, the most fascinating study subjects for students and researchers. Such a perspective is reflected in this book, where Jones outlines the different facets of primate behavior and shows that flexibility is a hallmark of primate behavioral patterns.

The concept of behavioral flexibility basically refers to the capacity to modify behavior in an adaptive way; or, as Jones defines it: behavioral flexibility represents a "toolbox of potential responses over time and space" (Chapter 9) allowing primates to adapt to heterogeneous biotic (including social) and physical environments. Because flexibility is a characteristic of different behaviors (such as foraging, mate choice, and dispersal) and may arise through a number of underlying mechanisms, it is crucial for any study of behavioral flexibility to clearly identify the specific behavior and to define how it is measured.

ix

In this book, Jones considers several detailed examples of how this could be achieved in primates. For instance in Chapter 3, male and female dispersal strategies in mantled howlers (i.e., different thresholds of when to disperse) are discussed as a response to changing habitat conditions. Long-term data showed that deforestation led to a rise in female but not male dispersal rates because female fitness is seen to be more influenced by resource availability than male fitness (females as "energy maximizers"). Another example of how to quantify behavioral flexibility is the "branch-break" behavior exhibited by male mantled howlers in Costa Rica (Chapter 6). This behavior is part of a compound display serving multiple purposes in mate choice and intrasexual competition depending on conditions of local competition. These examples suggest that future studies on behavioral flexibility should focus on situations in which the capacity for switching from one action to another is favored.

In other taxa such as birds and fish, research on behavioral flexibility has focused on different parameters such as the rate at which novel behavioral patterns occur (for a paper in primatology see, for example, Kummer and Goodall's 1985 article on "conditions of innovative behaviour in primates"). In Chapter 4 of Jones' book, this perspective is extended by discussing consequences of social cognition [referred to as the ability to perceive conspecifics as intentional agents—reflecting Kummer's (1971) definition of "social tool use"] as a generator for behavioral flexibility in the context of damaging and nondamaging behavioral tactics.

The history of research in primatology shows that studies have often benefited from applying concepts pioneered in other disciplines such as anthropology, psychology, and behavioral ecology. For instance, Crook and Gartlan's seminal paper in 1966 related differences in social organization of primate groups to differences in habitat and diet. Some ten years later, Clutton-Brock (1977) showed that within group differences are much more pronounced than differences between groups by analyzing the separate effects of several ecological variables on the same trait and using multivariate statistics. Another example is the use of experimental manipulation in the field to test hypotheses [e.g., Cheney and Seyfarth's (Seyfarth et al., 1980; Cheyney and Seyfarth, 1982) work on vervet communication], generally illustrating how research in primatology was advanced by shifting the predominant paradigm from rather descriptive to more analytical studies.

This book introduces several topics that have recently found widespread interest in behavioral ecology, for instance, the notion of inter- and intragenomic conflict. In Chapter 7, Jones sets out to explore the consequences of inter- and intragenomic conflict and their implications for primate behavior. It will be of particular interest to evaluate the significance of findings in behavioral genetics in other mammalian species with complex social systems for research conducted in primatology. Work on mice on the effects of so-called imprinted genes (genes that are expressed depending on the sex of the parent from which they were inherited) showed that these genes mediate a conflict between the sexes over maternal investment and have coevolved in an antagonistic way. For example, genes that are expressed only when inherited from the father serve to exploit maternal resource allocation during pregnancy and lactation by favoring higher maternal investment and positively influencing maternal behavior [i.e., females that carried a mutated gene showed impaired maternal behavior (see, for example, Li *et al.*, 1999)]. Studies on human mental disorders suggest that over and above their effects on provisioning behavior, imprinted genes significantly influence more complex social behavioral patterns in mammals (Isles and Wilkinson, 2000). To date, no one has assessed the potential implications of these results for research in primate behavior, despite their clear influence on mammalian parental behavior.

Given that these mechanisms operate in most mammalian species, would it be possible to interpret specific behaviors such as antagonistic male and female strategies as the result of intragenomic or intergenomic conflict? At present it appears that these findings may offer an explanation for a source of behavioral flexibility and underlying mechanisms. However, it will prove rather difficult to design experimental studies in primates that investigate such questions. A first step is taken in this book, where Jones outlines the implications for the antagonistic coevloution of primate male and female strategies such as coercive matings by males or female dominance and homosexuality. While the concept of antagonistic male and female strategies has been discussed in primatology for more than 20 years [e.g., Hrdy's work on infanticide (1979)], recent studies on other mammals suggest that there is clear evidence for such a process not only on the basis of behavioral observations but also as strategies of specific genes [e.g., Igf2 and Igf2R, (Haig and Graham, 1991)] favoring different optima of maternal investment for males and females. These remarkable findings may help to advance some of our previous interpretations of primate social behavior and view them in a new light of antagonistic genetic strategies.

Further, behavioral flexibility touches on other current hot topics in primatology. Flexibility in social behavior may play an important role in the evolution of cooperation in primates as demonstrated by recent studies in capuchin monkeys on inequity aversion. Brosnan and de Waal (2003) found that individuals will reject a reward for a task if they see a conspecific receiving a higher value reward for the same task. If increased behavioral flexibility results in higher sociality, as pointed out in Chapter 4, the ability to exhibit flexible behavior will have profound consequences for an individual's fitness. Work on baboons by Joan Silk and colleagues (2003) recently demonstrated that infants had a higher chance of survival if their mothers exhibited higher levels of sociality. These two examples illustrate that, on a larger scale, the study of behavioral flexibility in primates may improve our understanding of cooperation in nonhuman primates and humans.

While the advantages of behavioral flexibility seem intuitively obvious, it is more difficult to quantify and analyze the costs associated with increased flexibility. It is clear that greater flexibility will be selected for where it ultimately confers a fitness advantage to the individual. The difficulty arises because several behaviors might interact to create an overall more flexible phenotype, thus confounding a clear assignment of cost to a specific behavior. In addition, costs of increased behavioral flexibility might be incurred indirectly via a third parameter. For instance, costs associated with a higher dispersal rate in Japanese macaques may illustrate this point (Fukuda, 2004). Here, females (which are normally philopatric) had a lower threshold of dispersal in response to decreased food supply. This, in turn, caused increased male dispersal. As a result, not only was individual fitness negatively affected, but the species as a whole was put at a much higher risk of vulnerability because group sizes generally declined.

Other costs of increased flexibility are a higher predation risk, reduced rate of foraging (e.g., when trying novel food), time investment or energy required to exploit new resources. The identification and quantification of such costs, then, allows us to test the hypotheses about the adaptive value of behavioral flexibility. It may be argued that under conditions of limited food availability, individuals should be less willing to experiment and concentrate on food resources that will maintain the necessary rate of food intake.

Researching behavioral flexibility not only yields highly interesting results that help to elucidate the complexity of primate behavior but also has immediate applications in the conservation of primates and their habitats (Chapter 9). For example, analyzing the behavioral flexibility that is typical of different species in response to habitat change (which in turn affects food or territory availability) would help to better classify the vulnerability of a species to changing environments and, thus, to prioritize conservation efforts. A new additional approach in primate conservation seems particularly important because, as an example, the current classification of species into generalists (more flexible, thus, less vulnerable) vs. specialists has shortcomings as some specialist species may behave opportunistically if required. Thus, measures of the ability to exhibit flexibility in different behaviors (e.g., foraging, dispersal) will facilitate the estimation of how a given species will react to habitat disturbances such that a decrease in dispersal rates may be taken as an indicator for increased vulnerability.

Another reason to consider measures of behavioral flexibility in studies of primate biodiversity is the effect on rates of evolution in a species' genotype. Behavioral flexibility can be a major driving force for evolution because a high rate of novel behaviors may lead to increased evolutionary rates even if the variation in behavior is initially not due to underlying genetic variation but due to facultative responses that are culturally transmitted. The reason is that individuals who show high levels of behavioral flexibility will then be subject to different selection pressures. In turn, this will also affect whether newly arising genetic mutations will be selected for or against in a new context and may even lead to speciation.

The breadth of approaches to studying behavioral flexibility combined with examples of how to measure it in different behaviors make this book a valuable source for students and researchers alike. Jones shows that the knowledge of behavioral flexibility and the underlying mechanisms can significantly improve our understanding of primate social behavior and will help to unravel the complexity involved by identifying individual behaviors that, combined, result in phenotypic plasticity. Moreover, the study of behavioral flexibility holds particular importance for the conservation of primates and their habitats in directing the focus on behaviors that can be used as a measure for the adaptability to changing environments and the likelihood of a species' vulnerability.

Above all, this book is exciting to read for anyone who is interested in the scientific study of primate behavior. Jones introduces novel ideas into the field of primatology whose implications we only begin to grasp. I am looking forward to see future theoretical and empirical studies stimulated by the topics discussed here.

Reinmar Hager
Department of Zoology
University of Cambridge
Cambridge CB2 3EJ, UK

Preface

In a sense, I have never progressed beyond the age of six when children appear to be obsessed with the question, "Why?" Until very recently, "How?" questions failed to interest me very much, and they only captivate my imagination presently because I now see clearly in what ways the two sorts of questions are necessarily related (see Dixson, 1998, Glimcher, 2003, or West-Eberhard, 2003, if you are still unconvinced). I recall reciting a "book" to my maternal grandmother when I was three or four years old. She was ironing, and I was talking incessantly, as was my custom during my formative years. Even then, I was aware that I was privileged to have a caretaker who made me feel (and think!) that I was the center of her existence—more important than ironing, more important than cooking, more important than having an independent life of her own. This context, this privilege, afforded me the freedom to imagine a life beyond ironing and cooking, beyond the role of caring for others' dreams to become the reservoir of others' memories.

Wilson's (1971), *The Insect Societies*, influenced me during my years as a graduate student even more than Crook's (1964) monograph on weaverbirds, the seminal work in behavioral ecology. I believe that Wilson's book (especially Chapter 11) had a greater impact upon me because the comparisons between insects and vertebrates seemed so counterintuitive while the correlations described by Crook seemed, once demonstrated, self-evident. Wilson's 1971 volume and his *Sociobiology: The New Synthesis* (1975) helped me to see the study of social behavior as a *unified* enterprise, although recent work by Crespi and Choe (1997a; B.J. Crespi, personal communication) has helped me to understand the constraints on this approach.

Perhaps because of my own psychic particularities (e.g., temperament) combined with other factors (e.g., personal experience, education), Crook's (1970, p. xxvix) assertion that "co-operation ... seems often to be a subterfuge whereby an individual is enabled to gain or maintain that degree of social control of others at which his or her own behaviour is relatively unconstrained" influenced me in a manner that virtually insured my specializing in aspects of extreme selfishness, including spite, rather than, say, cooperation or altruism. As a result I have, possibly to a fault, studied topics such as interindividual conflicts of interest (e.g., power, homicide) and intraspecific social parasitism, including phenotypic manipulation. Early in my graduate training I formed a passionate interest in studying social behavior at the whole organism level. I was induced to study questions for their significance to ultimate causation primarily as a result of the excitement of seminars conducted by Jack Bradbury and Steve Emlen.

The primary goal of the present volume is to incorporate the extensive literature on behavioral flexibility in evolutionary biology and behavioral ecology into the canon of primatology in order to advance the Hamiltonian unification program within the Primate Order. My own conceptual framework, however, emphasizes the importance not only of optimization of genetic benefits but also of benefits to the phenotype, reflecting West-Eberhard's (2003) notion of the phenotype as a "bridge" between genotype and environment. This perspective appreciates that even where an individual's genotype is not directly favored by condition-dependent responses, promotion of the success of one's phenotype has the potential to enhance fitness over the long-term (see West-Eberhard, 1989, 2002).

West-Eberhard's (2003) subtle insights indicate that genotype and phenotype may be induced by different factors and may have different effects, often in interaction with each other and with the environment, suggesting that interests of the genotype and phenotype may be in conflict. In future, it will not be sufficient to think of responses simply in terms of their consequences for inclusive fitness. It will be necessary to consider individuals' "decisions" in terms of their effects on both genotype and phenotype. West-Eberhard's (2003) concept of the "bridging" phenotype and my extension of it in this book assume that the individual behaves in a manner that is fundamentally self-interested, thus, not to be confused with Wilson's (1980) discussion of "trait groups."

It is my hope that this project will promote the conceptual, theoretical, and empirical unification of primatology and the other (Darwinian) natural sciences. While most primatologists investigate behavior in relation to proximate (immediate) rather than ultimate (evolutionary) causation, and most primatologists have been trained to appreciate the significance of intraindividual variations in response, several recent publications have emphasized the need to integrate the proximate and ultimate perspectives (e.g., Dixson, 1998; Abbott et al., 1998; Jones and Agoramoorthy, 2003). The present volume places a primary emphasis upon ultimate causation since this conceptual, theoretical, and empirical canon has been effectively absent from the canon of the social sciences (compared to that of biology) until relatively recently. If, as Charnov (2002) puts it, organisms "live to reproduce," it may be a helpful exercise to explore how primate behavior and social organization are shaped when limited by the energetic and temporal constraints of reproductive effort.

It is also a goal of this volume to promote the experimental investigation of intraindividual phenotypic variation in primate behavior. Such a project would extend the success of experimental studies in behavioral ecology (see Piersma and Drent, 2003) and advance primatology as a truly hypothetico-deductive science with the potential to contribute to attempts to describe general principles of behavior. For example, the findings of Widdig *et al.* (2004), studying Rhesus macaques (*Macaca mulatta*), support recent attempts to develop a unifying theory of social interactions by lending support to the view that modified "tug of war" models of reproductive skew operate relative to taxon, environmental conditions, and other factors (see Hager, 2003).

Another example from the primate literature contributes to our understanding of "the mechanisms and processes that shape the expression of genetic variation in phenotypes" (Stearns, 2002, p. 10229). Through experiment, Maestripieri (2003) has recently demonstrated an apparent example of conserved traits between cross-fostered infant female Rhesus macaques and their biological mothers. In this study, infants raised by foster mothers demonstrated social and aggressive characters more similar to their biological than their adoptive mothers. This study raises the important point that plasticity may vary as a function of the target trait and that the relative advantages and disadvantages of developmental plasticity may differ as a function of age, sex and, possibly, other factors (e.g., nutritional status, kinship, or dominance rank). Maestripieri's research supports the view that intraindividual variation in response is induced by a genetic "switch" sensitive to the ecological and demographic environment (West-Eberhard, 1979; Gross, 1996; Jones and Agoramoorthy, 2003).

On the other hand, Weaver *et al.* (2004, p. 847) report that maternal behavior in rats "altered the offspring epigenome at a glucocorticoid receptor...gene promoter in the hippocampus." These results, which were reversed by cross-fostering, "show that an epigenomic state of a gene can be established through behavioral programming, and it is potentially reversible." These remarkable findings on "maternal effects" elucidate mechanisms of maternal behavior, in particular, "the nongenomic transmission of individual differences in stress reactivity across generations" (Weaver *et al.*, 2004, p. 847) and appear to contradict Maestripieri's (2003) study in which cross-fostered Rhesus macaques demonstrated affiliative and aggressive traits more similar to their biological than their adoptive mothers. If Maestripieri's results are confident, the contrast may indicate that some primates and, possibly, other taxa, demonstrate less plasticity in response to exogenously induced maternal stimuli (and, possibly, other exogenous stimuli).

Day *et al.* (2003) have recently published data which may facilitate the understanding of when individuals in populations or species will and when they will not exhibit behavioral flexibility. Seven callitrichid species from three genera (*Saguinus*, *Leontopithecus*, and *Callithrix*) were studied in their study of neophilia, innovation, and social attentiveness. Individuals belonging to species of *Leontopithecus* (lion tamarins) proved more neophobic than species in the other genera, consistent with the view that species dependent upon less manipulative and explorative foraging are more neophobic. Importantly, the work of these researchers did not support the view that dietary specialization was associated with neophobia. These findings may have broad implications for our understanding of the phenotype and exploitation of the niche.

Behavioral Flexibility in Primates: Causes and Consequences highlights similarities (signatures) and differences of primates to demonstrate that events in the world vary—one of the first lessons learned in introductory statistics. Species are subject to varying degrees of environmental heterogeneity, including stochasticity, a set of factors that may favor phenotypic plasticity, including, behavioral flexibility. Primates are among those taxa advanced to display an uncommon degree of behavioral flexibility (e.g., Boesch *et al.*, 2002), responses which, within the constraints of social parasitism, are presumed to optimize inclusive fitness for selfish (genetic and/or phenotypic) gain. Although many responses may be "the best of a bad job," the individual is expected to perform in his/her interests given the options available—all other things being equal.

The present book's discussion of behavioral flexibility in relation to evolutionary causes and consequences is advanced with full understanding that behavioral flexibility can only be demonstrated to be adaptive if evidence, preferably experimental, is presented showing that the relevant features have been shaped by evolution to enhance lifetime reproductive success (fitness, Hamilton, 1964; West *et al.*, 2001; Reeve, 2002). Stressing the importance of the study of behavioral flexibility to this program, Piersma and Drent (2003) state: "Rather than emphasizing that a capacity for phenotypic change is an adaptation (which it might well be: Pigliucci, 2001, p. 231), we argue that intra-individual trait variation itself should be used to evaluate the 'goodness of design' criterion for phenotypic adaptation (Williams, 1966)." For example, Garber and Leigh (1997) illustrate how comparisons in patterns of reproduction and infant care among small-bodied Neotropical primates yield strong inferences about function, in particular the energetic costs of reproduction and maturation, factors fundamental to an understanding of life history tactics and strategies.

Time and energy are limiting for organisms, making the allocation of these resources essential to an understanding of life history (e.g., Schoener, 1971; Charnov, 2002). Furthermore, every act performed by an individual will have effects with potentially significant consequences for the fitness of the actor and all individuals influenced by the act(s) (West *et al.*, 2002). For this reason, it is important to describe events before and subsequent to any response(s) of interest (e.g., Jones, 1983).

I am an advocate of the expanded use of mathematical models and other quantitative techniques not often employed in primatology, an inclination reflected in the present bibliography. While this project has received generous encouragement from numerous colleagues within and outside primatology, many of my peers remain skeptical, resistant, or, even, hostile to theory as defined in Economics, Ecology, and Evolutionary Biology. Numerous primatologists have pointed out that mathematical treatments can help the scientist to frame his/her questions clearly and precisely. However, perhaps the most important utility of mathematical models, simplified as they may be, is to test whether our questions conform to particular assumptions, such as those of Darwinian theory. It seems insufficient to collect data and to ask questions without reference to a conceptual framework that has received theoretical support. Theoretical treatments, then, can provide initial stages in scientific programs in order to constrain our search relative to the questions worth asking and to guard against what Wynne (2004) has called "the perils of anthropomorphism." Few behavioral and social scientists would object to using a chi-square "goodness-of-fit" test to assess their hypotheses before investing in expensive empirical investigations. Utilizing other "goodness-of-fit," estimation, and simulation techniques have similar and, in many cases, more powerful utility (see Dunbar, 2002).

The present volume proposes that genetic conflicts of interest are ubiquitous in primates who may employ force, coercion, persuasion, persistence, scrambles, cooperation, altruism, exploitation, manipulation, social parasitism, dispersal, or spite to resolve or to manage them. Where one individual or group imposes severe costs to inclusive fitness upon a conspecific, the latter may adopt a counterstrategy in an attempt to minimize its costs. This counterstrategy may, in turn, impose costs upon the original actor(s), and so on, possibly yielding an evolutionary "chase" ("interlocus contest evolution", Rice, 2000; Nunn, 2003). The evolution of phenotypic plasticity and/or behavioral flexibility in primates may often pertain to attempts to mitigate genetic conflicts of interest, and, consistent with Trivers' (1972) treatment of parental investment, Schoener's (1971) classic paper leads to the conclusion that for females ("energy-maximizers"), conflict will pertain primarily to competition for food (that can be converted to gametes and/or offspring) while, for males ("time-minimizers"), conflict will pertain primarily to competition for mates, *ceteris paribus*. Recent empirical work on primates supports Schoener's theoretical formulations (e.g., male *Pan troglodytes*, Newton-Fisher, 2002; male and female *Alouatta palliata*, Clarke and Glander, 2004).

While mutation or other intragenomic effects may be a source of intraindividual and interindividual variation in the behavior of primates, most behavioral flexibility in social taxa of this order probably arises via trial and error, associative, or cognitive processes as novel, facultative responses to heterogeneity in physical and/or biotic, including social, regimes. The present volume explores the contexts, ecology, development, and evolution of condition-dependent responses in primates. These responses will sometimes be manifestations of tactics and strategies to optimize survival, lifetime reproductive, and/or phenotypic success, and/or may be counteradaptations to intraspecific social parasitism and intergenomic conflict.

It is my hope that students, specialists, and the general public interested in the diverse manifestations of environmental heterogeneity upon primate behavior and social organization will find this book a "good read" and a rich source of ideas for discussion as well as theoretical and empirical investigation. Environments are often unpredictable and the outcomes of individual decisions often uncertain, costly, and, possibly, risky—conditions that will limit the ability of individuals to behave optimally. A broader understanding of these states and their outcomes than is currently reflected in the primate literature has the potential to generate a revised view of the landscapes upon which primates behave and evolve and the ways that preadaptations, adaptations, and (genetically autocorrelated) responses to these regimes (see Miller, 1956; Lewontin, 1957; Slobodkin and Rapoport, 1974; West-Eberhard, 2003) may have favored the characteristics commonly associated with human beings (see, for example, Miller, 1997).

Acknowledgments

In nonhuman primate societies, acknowledgments would take the form of rear-present postures, kissing, lip-smacking, G-G rubbing, grooming, or, possibly, greeting ceremonies. Despite differences, for most primates, including humans, signs of appreciation are, in part, a reflection of our inherent social natures and the extent to which the expression of our individuality is dependent upon the acts of others. The spirit of gratitude with which I write these acknowledgments, then, is informed by the motivating principle of the book as a whole—the biotic (including social) and abiotic constraints giving rise to homologies, convergences, and conservative effects within and between populations and species in their expressions of interindividual behavior and socio-reproductive organization.

My career can be described as the outcome of a multitude of combinatorial effects—abiotic (class, marital satisfaction) and biotic (gender, race). Numerous persons have contributed to the production of this book, some directly, many others indirectly. My interest in the evolution of social behavior would not have matured without the patience, wisdom, and knowledge of numerous professors, colleagues, and fellow students and post-docs. I am grateful, in particular, to Jack Bradbury, Bill Dilger, Steve Emlen, Dick Lewontin, Bob Trivers, Mary Jane West-Eberhard, and the late Jasper Loftus-Hills who have served to model the very best practices of ethology, sociobiology, and behavioral ecology. Irwin Bernstein, Ruth Buskirk, Irenäeus Eibl-Eibesfeldt, Sarah Hrdy, the late Grif Ewer, and many others have also provided helpful input and constructive criticism. My participation in the course, Tropical Ecology 73-2, sponsored by the Organization of Tropical Studies, was a critical personal and professional influence, and I am especially devoted to the training provided by Don Wilson, Norm Scott, Dan Janzen, Dick Root, Mary Willson, Jeff Graham, Gordon Frankie, and José Sarukhán. In some ways, the most important early influences were Bill Dilger, Bob Johnston, and Fred Stollnitz, biopsychologists responsible for introducing me to the discipline of Psychology (the science of behavior, brain processes, and emotions) and the specialization of Primatology in a manner that was inclusive of several paradigms and research programs for the conceptualization and investigation of behavior, individual decision-making, and social organization. Finally, I gratefully acknowledge the financial support of the Ford Foundation and National Research Council during my doctoral and postdoctoral training, respectively.

Throughout my career, numerous books and journal articles have proven particularly influential. However, John Hurrell Crook's 1964 monograph on weaver birds, Schoener's (1971) paper, "Theory of feeding strategies," E.O. Wilson's 1971 book, *The Insect Societies,* Dick Lewontin's 1957 paper, "The adaptations of populations to varying environments," Bob Trivers' 1972 chapter on parental investment and sexual selection, and Mary Jane West-Eberhard's 1979 paper, "Sexual selection, social competition, and evolution," deserve special mention as the first academic works to "take my breath away" upon first reading. Odum's (1971), Emlen's (1973), and Roughgarden's (1979) ecology textbooks, with their functional approaches, provided primary imprinting to that field. Reinforcing the emphasis upon function, Jack Bradbury, Steve Emlen, and Dick Lewontin are responsible for impressing upon me the significance to organismal responsiveness of environmental heterogeneity, a conceptual framework that has guided all of my thinking about social behavior. Jack Bradbury and Steve Emlen taught me to respect the power of time and energy, and Dick Lewontin taught me to respect the power of stochasticity.

The present volume would not have been produced in a timely manner without the technical and other assistance of Sarah DiGloria who produced all of the line drawings, edited several of the photographs, compiled the final document before submission to Springer, and, most importantly, prepared the bibliography. While I am responsible for any errors in the final product, Sarah's enthusiasm and support for this project provided the necessary stimulus whenever my own attention to the program waned, thereby insuring that this book remained a priority for the duration of its preparation. Russell H. Tuttle, Editor of Springer's Developments in Primatology Book Series, and Andrea Macaluso, Senior Editor of Life Sciences at Springer provided oversight during all stages of the project. Krista Zimmer, Macaluso's Editorial Assistant, artfully managed my numerous queries as well as the detailed paperwork associated with the book's preparation. Robert Maged, Senior Production Editor at Springer, expertly saw the book through its final phases. Very helpful comments by two anonymous reviewers significantly improved the manuscript.

The aesthetic value of the present volume has been enhanced by the generous donation of photographs by Carla Boe, Fumio Fukuda, Josh Palmer, Mike Seres, and Julie Ann Wieczkowski. Sarah DiGloria and the late Jack Daniel edited my howler photos, as well as several other images, with characteristic expertise. Publishers who provided permission to reprint original figures or to base my own figures upon published figures are gratefully acknowledged in the figure legends. I also want to thank certain individuals at Livingstone College, in particular, librarian, Matyas Becvarov, for obtaining several critical bibliographic references and my teaching colleagues, Bob Williams, political scientist, and Walter Ellis, a professor of social work, for engaging intellectual interactions, personal and professional support, and an unflagging willingness to discuss nonhuman primates even though their own research pertains exclusively to humans.

I am very grateful to Reinmar Hager for agreeing to write the Foreword for the present volume. His comprehensive understanding of sociobiology and behavioral ecology—both theoretical and empirical—represents the new wave of young researchers investigating primates and other social animals, including humans. Definitions in the text are based primarily upon West-Eberhard (2003), Piersma and Drent (2003), and Choe and Crespi (1997) and are presented in a Glossary sited before the bibliography, which I hope readers will find useful. Taxonomy is based on Groves (2001). Bernard J. Crespi graciously allowed me to read a very stimulating unpublished manuscript on the evolution of social behavior. Electronic communication with Fumio Fukuda inspired me to write Chapter 2. Robert Poulin provided a reference and constructive criticism for Chapter 3, which significantly improved its presentation. Reinmar Hager offered constructive criticism on Chapter 4, and Glenn Hausfater, Paul Hertz, Tim Caro, Jerry Brown, Gordon Orians, and Jeanne Altmann commented on elements of this chapter as presented in an earlier, unpublished manuscript. Also pertaining to Chapter 4, Josep Call suggested citations that proved useful, and Luke Jones provided wise advice on a figure originally included in this section of the book.

Dick Lewontin, E.O. Wilson, Mary Jane West-Eberhard, Irwin Bernstein, Bill Dilger, and Ken Weber graciously read and made helpful comments on my work on temporal division of labor in mantled howler females during my year as a post-doctoral fellow at Harvard, and Anthony Rylands permitted me to use figures and text from Jones (1996a) as components of Chapter 5. Chapter 6 was developed with the expert input of Robert J.H. Payne. Stuart Altmann, Irwin Bernstein, Paul Garber, Ryne Palombit, and Joseph Soltis also provided helpful critical input on an earlier version of this chapter, and Katherine Jack, Joe Manson, Leann Nash, Ryne Palombit, R.J.H. Payne, and Joseph Soltis provided useful references. Rob Horwich kindly shared his observations of *Ateles* in El Salvador which enhanced my understanding of "branch-break" behavior in these monkeys, and Rob deserves special recognition for providing me with the opportunity to study *Alouatta pigra* in Belize. Reinmar Hager's very knowledgeable and helpful comments and suggestions on genomic conflict improved my understanding and presentation of these mechanisms discussed in Chapter 7. I am grateful to Wendy Saltzman for assisting me with citations on callitrichids and for the resources of the Wisconsin Primate Research Center and its former librarian, Larry Jacobsen. The tools, Primate Lit and the Primate-Science List Serve, have proven to be important sources of information throughout the preparation of this book. One correspondent, through the listserve, Inês Canavarro Morais, generously provided a critical quotation and citation.

My remaining, and more personal, acknowledgments are extended to my late mother, Clara K. Jackson Brown for providing early enthusiasm for and training in the study of mathematics and science and in the care and management of plants and animals. The late Jasper Loftus-Hills introduced me to the study of social parasitism, an interest that had been ignited by the exhortations of my graduate advisor, Bill Dilger, to study the topic of mimicry in primates. In no small measure, my persistence as a student of behavior and social organization and my willingness to consider primate (including human) behavior from the perspective of the Hamiltonian unification project is a result of Jasper's friendship, tutelage, and uncommonly creative style of biological thinking. Jasper's untimely death created a significant loss. Nonetheless, my personal life continues to be enhanced to an immeasurable degree by the ongoing concern, good humor, and good sense of my children—Dalton Anthony, Julie Karin, and Miguel Luke. Their apparent ability to understand and tolerate my passionate and frequently undivided devotion to my work has no doubt contributed in some measure to their own highly developed "emotional intelligence" and career success. Finally, this volume is gratefully dedicated to my late maternal grandmother, Clara Kersey Jackson, whose selfless investments of time and energy during the first five years of my life have been very helpful over the long term.

Introduction to Intraindividual Variation of Primate Behavior

The behavior of populations is an emergent property of the reactions of individuals to their circumstances...

Macdonald and Johnson (2001, p. 367)

Introduction

Events in the world may vary simply by chance alone. Thus, caution must be employed where attempts are made to generalize. As Darwin (1859, 1871) understood, it is through the study of diversity that one identifies general patterns, most often by descriptive approaches prior to the application of appropriate quantitative methods of inference. Numerous students of primates have sought to describe broad patterns of response within the Order (Smuts *et al.*, 1987; Dunbar, 1988; Dixson, 1998; Box, 1991; van Schaik and Kappeler, 1997; Muller and Thalmann, 2000; Kappeler and Pereira, 2003; Jones, 2002a; Jones and Agoramoorthy, 2003; Maestripieri, 2003a; Fewell, 2003), and monkeys and apes have long held a fascination for humans because of phylogenetic proximity. For this reason, primates of the Old World (Africa and Asia) are relatively well known compared to primates of the New World who diverged earlier from the anthropoid line.

Primates evolved from primitive insectivores, later diverging toward a fruit- or leaf-eating mode with concomitant changes in dentition and digestion. New World and Old World primates arose from an insectivorous common ancestor, and some extant species occupy niches similar to those of Paleotropical species (e.g., squirrel monkeys and talapoins; howler monkeys and langurs; spider monkeys and chimpanzees). Arboreal life is a primitive trait in the Order, a condition favoring stereoscopic vision, a grasping hand, impressive muscular coordination, and relative brain enlargement. Despite these traits, evolutionary proximity to our own species, and aesthetic value, nonhuman primates have no known "key" roles in ecosystems, possibly because the distribution and abundance of their prey are almost certainly determined by factors other than primate predation.

In addition to phylogenetic proximity to *Homo sapiens*, nonhuman primates are important targets of research in their own right as well as for the documentation of general patterns and principles of primate and mammalian biology. Indeed, since it is through the investigation of diversity that general patterns and principles can be identified, one of the goals of the present book is to evaluate the causes and consequences of behavioral, including social, flexibility of primates, including humans. As pointed out in the Foreword and Preface to this volume, most primates, including humans, are noted for their behavioral plasticity. The broad literature on this topic from evolutionary biology and behavioral ecology has not been integrated into the primate literature. One of the goals of this book is to discuss primates in relation to these theoretical and empirical treatments. I also hope to show that there are *signatures* (diagnostic features) of behavioral flexibility in primates worthy of intense investigation and that these features, once studied within a Darwinian paradigm, will facilitate the conceptual unification of primatology with the other natural sciences.

Concepts and Definitions Related to the Study of Behavioral Flexibility

One of the driving concepts of the present volume is that of *motivation*, a proximate construct that some have abandoned because of its empirical inconstancy. To the contrary, I have had a solid faith in the utility of motivation as a construct since discussing it with Robert Hinde in Bob Johnston's undergraduate animal behavior course at Cornell in the late 1960s. It seemed evident to me after that experience that the significance (meaning in terms of function) of behaviors might be analyzed in a manner similar to the "phylogenesis" of genetics, development, morphology, and behavior (Henikoff, 2003). It was not until working with Irenäus Eibl-Eibesfeldt at Seewiesen in 1980, however, that it became clear to me that motivation could be assessed quantitatively by studying *sequences* of behavior (Jones, 1983a; Emlen and Wrege, 1994; Grafen, 2002), using appropriate quantitative techniques to detect patterns of response between actor and recipient before, during, and as a consequence of simple or compound behaviors and response sets. The patterns of response identified in Chapter 3 as primate signatures, amenable to sequence analysis, are of import in part for what they reveal about condition- and situation-dependent motivation relative to environmental regimes.

Proximate causation is intimately linked with ultimate (evolutionary) causation, and I hope to influence the reader to think of environmental heterogeneity as creating opportunities for individuals to respond in novel ways to challenges afforded to survival, reproductive success, and the phenotype. Individuals may "decide" (consciously or otherwise) to respond differentially within and between situations. Helms Cahan *et al.* (2002) point out, for example, that where an organism "decides" not to cooperate, (s)he may either disperse (Chapter 2) or remain in his/her group as a social parasite (Chapter 3). While individuals responding to temporal and spatial heterogeneity may be motivated (not necessarily consciously) by different proximate goals (e.g., the reduction of unpredictability, uncertainty, error, or risk), their ultimate (evolutionary) goal remains the same as animals (including humans) living in less heterogeneous conditions—the optimization of inclusive fitness and benefits to the phenotype. The present volume, then, hopes to advance the Hamiltonian unification program (Hamilton, 1964); although, as discussed in Chapter 3 of this volume, certain responses by conspecifics, in particular, socially parasitic ones, have the potential to derail an individual's self-interested efforts.

Behavior is that component of the individual most directly exposed to the external environment, to environmental perturbations, and potentially, to selective pressures (Mayr, 1963; Manley, 1985; West-Eberhard, 2003). Early work on behavior was heavily influenced by the *genetic* approach (Craig and Guhl, 1969; Zuckerman, 1932; Eibl-Eibesfeldt, 1970). In this view, allozyme markers commit the individual to a particular strategy or mix of strategies within an environmental mosaic and are not modifiable within the lifetime of the individual. Subsequent research emphasized the importance of *physiological* (Watson and Moss, 1970; Bernstein *et al.*, 1974; Vandenbergh, 1983) and *behavioral* (Deag and Crook, 1971; Staddon, 1983; Mazur, 2002) mechanisms as *epigenetic* processes of homeostatis for local adaptation to short-term environmental fluctuations.

In this book, *behavior* is understood to imply any neuromuscular activity exposed, directly or indirectly, to the selective environment, including associated physiological mechanisms and pathways (e.g., emotional, cognitive). This restrictive definition, then, does not consider emotions and brain processes, per se, *behaviors*, unless they comprise some component of the phenotypic space or surface subject to evolutionary processes. These characters will typically be continuously varying and polygenetic quantitative elements whose analysis at the molecular and genotypic levels of organization (e.g., "sociogenomics": Lim *et al.,* 2004) is beyond the scope of this book. My notion of *phenotype*, in particular, the *behavioral phenotype*, and its relationship to development, however, is derived from recent treatments (Scheiner, 1993; Kirschner and Gerhart, 1998; Schlichting and Pigliucci, 1998; Pigliucci, 2001; Debat and David, 2001; West-Eberhard, 2003; Piersma and Drent, 2003; Bolnick *et al.*, 2003; Francis *et al.*, 2003; Crabbe and Phillips, 2003; Jones and Agoramoorthy, 2003; Spitze and Sadler, 1996; Reader and Macdonald, 2003) and the related literature which preceded them (Lewontin, 1957, 1974, 2000; Levins, 1968; Levin, 1976; West-Eberhard, 1979, 1989), viewing the phenotype as an expression of "dynamic" environmental agents interacting with "epigenetic pathways" (West-Eberhard, 2003).

The focus of this book, *behavioral flexibility* (after Piersma and Drent, 2003, p. 228, Table 1), sometimes termed "behavioral plasticity" and a component of phenotypic flexibility, implies reversible intraindividual (single-genotype) changes in behavior in response to an environmental (situational) stimulus or stimulus array detectable by the organism. The stimulus or stimulus array may be induced endogenously (e.g., through an endocrine response; Ziegler *et al.*, 1987) or exogenously (by a detectable change in temperature or food supply; Melo *et al.*, 2003; Keith-Lucas *et al.*, 1999) or by a combination of these two sources (e.g., Kraus *et al.*, 1999; Ziegler *et al.*, 2004). Crespi and Choe (1997a, p. 506) point out that, for organisms characterized by overlap of generations (e.g., all primates), behavioral flexibility should be favored since differential (genetic and/or phenotypic) benefits (tradeoffs, optima) from alternative behavioral tactics and strategies should change significantly over time.

Behavioral flexibility may occur in the form of a "facultative," condition-dependent, and/or compensatory response resulting from a genetically and/or physiologically induced "switch" or may be induced by a developmental program ("developmental plasticity") sensitive to competitive regimes and changes in environmental stimuli (Gross, 1996; Keller, 2003; West-Eberhard, 1979; 1989; 2003). Behavioral flexibility may also occur as responses expressed as a result of "trial and error," associative, and/or cognitive processes, mechanisms that may not be genetically or physiologically induced. These operations are usually assumed to be induced by abiotic and biotic factors external to the organism (e.g., Mazur, 2002; Maestripieri, 2003a). On the other hand, some major reviews of psychological processes (e.g., Silva *et al.*, 1997), which are generally thought to represent flexible responses, explain these phenomena (e.g., learning) from the perspective of genetics, suggesting that the genome is an important conceptual framework for many psychologists, particularly comparative and physiological psychologists and neuroscientists. In the primate literature, use of the term "behavioral flexibility" often implies that a response is a facultative (reversible) one (e.g., Kowalewski and Zunino, 2004) and is an important operation to study because its potential is thought to have evolved in response to environmental heterogeneity, benefiting individual survival, reproductive success, and/or the phenotype in these regimes. Behavioral flexibility may also expose components of the phenotype with underlying genetic variation to selection (West-Eberhard, 1989). As a consequence of behavioral flexibility, variation in the between-individual component of the behavioral phenotype will increase, with important consequences for competitive regimes within populations, a topic addressed in Chapter 2 of this volume. For most of the examples of behavioral flexibility discussed in this book, the precise regulatory mechanisms will generally be unknown.

Where behavioral flexibility is favored, increased between-individual variation can be achieved either by a "true generalist" strategy in which the individual performs as, in effect, a "Jack of all trades," exhibiting a variety of functions or as a "polyspecialist," exhibiting a narrow range of relatively discrete tasks which may be associated with alternative responses or other traits (e.g., morphological features, search images: West-Eberhard, 2003, p. 382). A generalist or polyspecialist strategy permits the individual the option of narrowing or broadening their range of response(s) to the environment with consequent narrowing or broadening of the individual's utilization of niche space relative to its conspecifics utilizing the same environmental components. A narrowed response repertoire by a generalist, however, would not be expected to be as efficient as the range of response by a specialist in the same conditions. Behavioral flexibility will permit the individual to exploit a narrower or wider range of situations in the abiotic and/or biotic (including social) environment, depending upon the response(s) most likely to optimize inclusive fitness and/or phenotypic success. Table 1.1 displays hypothesized origins and examples of primate behavioral flexibility and plasticity (based in part upon West-Eberhard, 2003) from the literature, both unselected (e.g., before selection) and after selection, enhancing phenotypic plasticity.

The Costs and Benefits of Behavioral Flexibility

The abiotic or biotic (including social) environment may change in a predictable or an unpredictable manner relative to the individual's "preparedness" as a result of genotype and/or prior experience. In such situations, the individual (consciously or otherwise) may "decide" upon a response or response array likely to maximize benefits and minimize costs to individual genotype and/or phenotype. As an example for primates of this optimality approach to cognition, Bales *et al.* (2001) showed that cooperatively breeding female golden lion tamarins (*Leontopithecus rosalia*) in the wild base "decisions" about the size of litters in part upon the number of helpers available to them. Assumptions about optimal decision-making are central to contemporary evolutionary thought (Hamilton, 1964; Trivers, 1985) and to many studies conducted in primatology (e.g., Smuts, 1985). In order to optimize inclusive fitness and minimize error, in order to exert as much control as possible over uncertain conditions, and/or in order to "track" environmental changes (see, for example, Jones, 1997b), behavioral flexibility may benefit the individual by counteracting or minimizing the potentially deleterious consequences of heterogeneous conditions, by fine-tuning the individual's responses to the situation, and/or by creating novel responses to heterogeneous circumstances.

In addition to humans, numerous primate species are thought to display a significant degree of behavioral flexibility (e.g., callitrichids: Goldizen,

Table 1.1. Hypothesized Origins of Novelty Generating Behavioral Flexibility (Reversible Intraindividual Behavioral Variation; Piersma and Drent, 2003) in Primates (Based in Part upon West-Eberhard, 2003), Including Examples from the Primate or Other Mammalian Literature.

Unselected or before selection[a]	Possible example(s) in primates or other vertebrates
Mutation	Scriver and Waters, 1999; Bjedov *et al.*, 2003
Epigenetic effects	*Macaca* (Goodhill *et al.*, 1997)
Parasites and/or Disease	*Homo sapiens* (Martinez *et al.*, 2001)
Learning (all mechanisms, including association, conditioning, copying, matching, mimicry, social parasitism, and cognitive processes)	*H. sapiens* (DeFries *et al.*, 1986; Jones, 1986); *A. palliata* (mimicry, for example, paedomorphic vocalizations: Jones, 1980, 1985a, 1997a); Primates (Jones, 2003a)
Position effects (e.g., translocation)	Dillon, 2003
After selection[b]	
Different thresholds to different stimuli	*H. sapiens* (Hewitt and Turner, 1995)
Genetic polymorphism	*H. sapiens* (Halushka *et al.*, 1999)
Generalist phenotype[c]	Females compared to males in the same conditions (?)
Polyspecialist phenotype[d]	Males compared to females in the same conditions (?)
Duplication	*H. sapiens* (Wells and Warren, 1998; Margolis *et al.*, 1999)
Deletion	Birds and primates (Sundstrom *et al.*, 2003)
Reversion	*M. mulatta, Papio anubis* (Macy *et al.*, 2000)
Heterochrony	*H. sapiens* (Godfrey and Sutherland, 1996); *M. nemestrina* (German *et al.*, 1994)
Heterotopy	Rats (Maly and Sasse, 1987)
Cross-sexual transfer	Alternative reproductive tactics and strategies in primates (Vasey, 2002; see Jones and Agoramoorthy, 2003)
Quantitative shifts and correlated change	Rats, marmosets, and humans (Dumont *et al.*, 2000)
Combinatorial evolution at the molecular level ("genomic introns")	Primates (Subramanian and Kumar, 2003; Yi *et al.*, 2002)

NB: Behavioral flexibility may be favored by heterogeneity in abiotic (e.g., temperature, humidity) or biotic (e.g., nutrition, fatigue, interaction rates) regimes, including conditions within the organism. Each hypothesized origin (mechanism) implies organization and/or reorganization of some component of the organism (e.g., the behavioral phenotype)

[a] A response must be correlated with genetic variation in order for it to be acted upon by selection (Van Tiendren and Koelewijn, 1994). Not all responses that reorganize the phenotype, then (e.g., some learned responses), will have evolutionary potential (see West-Eberhard, 1989; Schlichting and Pigliucci, 1998).

[b] With the possible exception of "different thresholds to different stimuli," these mechanisms are expected to be regulated by a "switch" mechanism (see West-Eberhard, 1979, 2003; Gross, 1996; Jones and Agoramoorthy, 2003). This treatment assumes that plasticity represents a trait with underlying genetic variability that can be acted upon by selection (see Piersma and Drent, 2003).

[c] West-Eberhard (2003, p. 382) defines a "generalist" as an organism who "performs a broad and highly variable range of tasks, often with little distinction in the morphology, sensory capacities, and behaviors used to accomplish each."

[d] West-Eberhard (2003, p. 382) defines a "polyspecialist" as an organism who "performs a limited number of distinctive alternative tasks using alternative sets of behaviors, morphological equipment, search images, and so forth."

1987; Saltzman, 2003; howler monkeys: Crockett and Eisenberg 1987; Crockett 1998; Jones, 1999a, 2000; capuchins: Fragaszy *et al.*, 1990; Fragaszy and Perry, 2003a; cercopithecines: Swedell, 2002; Kamilar, 2003; orangutans: van Schaik *et al.*, 2003; chimpanzees and bonobos: Boesch *et al.*, 2002; humans: Hrdy, 1999a; Gehring and Willoughby, 2002). For any analysis of behavioral flexibility in response to temporal or spatial environmental heterogeneity, it is important to assess the extent to which environmental or situational change may have a net positive impact upon inclusive fitness. Behavioral flexibility, if not initially deleterious, will only be beneficial to the individual up to some level of cost where it becomes deleterious to inclusive fitness (Relyea, 2002; Whitlock, 1996; Kawecki, 1994; Pallier *et al.*, 1997). Costs may increase, for example, if the likelihood(s) of behavioral error or inaccuracy increase with an increase in behavioral flexibility (see Chapter 9) or if the expression of behavioral flexibility represents a waste of time and energy. The investigation of costs is of particular import since some threshold of condition-dependent costs is expected to expose the individual to selection, assuming underlying, correlated genetic variation.

Components of Phenotypic Flexibility, Including Behavioral Flexibility

Most models of phenotypic flexibility, particularly quantitative genetic models, have considered intraindividual ("within-individual") variation to represent *noise* (random variations of stimuli and/or responses) whose incorporation would decrease a model's predictive value (Piersma and Drent, 2003; West-Eberhard, 2003; Kosslyn *et al.*, 2002). Piersma and Drent (2003) show, however, that phenotypic flexibility, in our case its behavioral component, can be reduced to a component of total phenotypic variance that is *reversible* and a component that is *nonreversible*. As stated above, it is the *reversible* component that is the focus of the present volume, and this component may have particular import for the evolution of primate behavior (Jones and Agoramoorthy, 2003; Miller, 1997). Piersma and Drent (2003) point out that the above division into reversible and nonreversible elements permits their separate quantification, an assessment of their relative contribution to total phenotypic variance, and also their *interaction*. These authors show that the interaction term reveals the dependence of reversible variations upon stage of development (i.e., the irreversible component). Piersma and Drent (2003) argue, further, that length of development is expected to be negatively correlated with phenotypic flexibility. Thus, older individuals are expected to exhibit a lower degree of behavioral flexibility than younger individuals (see Jones, 1996a; Parthasarathy, 2002; Palleroni and Hauser, 2003). The elegant analysis of Piersma and Drent (2003, p. 231) demonstrates that intraindividual variation has three components: "a genetic component and two environmental components (a reversible and an irreversible one)."

Members of the Order Primates are good candidates for investigation of behavioral flexibility since their responses are relatively generalized (Vaughan, 1978; Fleagle, 1999; Kitchen and Packer, 1999; Boesch *et al.*, 2002; Padilla and Adolph, 1996), yielding more or less totipotent phenotypes capable of performing a variety of tasks and occupying relatively broad niches, all other things being equal. Crespi and Choe (1997a, p. 514) point out that most social vertebrates are characterized by totipotency, contrasting these species with those exhibiting castes. The evolutionary trajectories of primates have led to suites of traits subject to flexible combination and recombination, optimizing fitness and, potentially, leading to the construction of novel communication elements.

In the primate literature, the most frequent examples of behavioral flexibility pertain to foraging tactics and strategies, communication signals and displays, social context, and tool use. Recent research has investigated behavioral flexibility as manifested in infanticidal behavior (see Saltzman, 2003), alternative reproductive tactics and strategies (e.g., Jones and Agoramoorthy, 2003; Dunbar, 1982), and dominance rank (Jack, 2003). Studying Barbary macaques (*Macaca sylvanus*), for example, Ménard (2002) discussed flexibility in choice of dietary items, particularly between seasons. This author identified two feeding "phases," a "granivorous phase" occurring when herbaceous seeds and/or acorns predominated and an "insectivorous phase" observed when this proteinaceous food source was most abundant. Numerous other primatologists have also reported adjustments in feeding behavior as a result of changing food conditions and/or individual requirements (e.g., Glander, 1975). In folivorous primate species such as the *Alouatta palliata* (mantled howling monkeys) studied by Glander, animals "switch" to mature leaves when preferred food (e.g., fruit, flowers, or new leaves) is scarce. In all of these studies, switching from one food source to another may occur within and between seasons, and patterns of food selection may differ by age/sex category as well as individually (see Glander, 1975: Altmann, 1998). Most of these responses appear to represent facultative responses to changing conditions in the environment external to primates, although individuals may also alter their food choices in response to endogenous factors such as dietary deficiencies (e.g., Krishnamani and Mahaney, 2000).

As referenced in Hager's Foreword to this book, many studies of primate communication have investigated responses to potential predators (e.g., Oda, 1998). Research on mother:infant communication has also been common (e.g., Nunn, 2000). The large literature on primate communication, especially vocal communication, documents a significant degree of flexibility in these signals, which may be influenced by a broad range of endogenous (e.g., individual motivation) and exogenous (e.g., social) factors. Studying cotton-top tamarins (*Saguinus oedipus*: see cover photo), Rousch and Snowdon (1999) documented that social status influenced the development of adult types and usage of food-associated calls. Although the authors did not speculate about the possible mechanisms controlling their findings, these results suggest that cooperatively breeding callitrichids may be endowed with a genetically induced "switch" mechanism for three separate developmental processes governing calling behavior. This study further suggests that the "switch" mechanism may be sensitive to thresholds such as time since separation from their natal group.

As evidence of the long history of study on tool use in vertebrates, including primates, van Lawick-Goodall's 1970 chapter documented this highly variable response in a broad range of species, which is generally thought to result from social learning. The investigation of tool use in the primate literature typically describes these behaviors in Paleotropical species. In 2001 and 2002, Humle and Matsuzawa reported their remarkable observations of behavioral flexibility in material food culture of *Pan troglodytes verus*, particularly nut cracking and ant-dipping techniques. These authors point out that these responses may be a function of abiotic or biotic factors or social ones. In another important publication on behavioral flexibility and evolution in *Pan*, Boesch *et al.* (2002) present chapters documenting behavioral diversity in chimpanzees and bonobos from a variety of sites in Africa. In this volume, several reports exhibit ecological differences in tool use that may be associated with differential degrees of genetic variability between sites. While the study of tool use in Neotropical primates is increasing (e.g., Fragaszy and Perry, 2003), additional research is required to investigate the potential for behavioral flexibility in material culture by Platyrrhines (see McGrew, 1998).

It is the consensus of ecologists that phenotypic flexibility has arisen in response to temporal and spatial environmental heterogeneity (variation in the environment; Piersma and Drent, 2003), and this book intends to explore the causes and consequences of environmental heterogeneity for primates. The generally conservative nature of evolution (e.g., Maestripieri, 2003b) leads to the expectation that the mechanisms and functions discussed in this book will apply, as well, to other groups of organisms, particularly social mammals, facing similar environmental regimes. The *caveats* and analyses of Brooks and McLennan (2002) are acknowledged, however, and phylogeny is expected to constrain (by imposing differential costs or benefits) the evolution of behavioral flexibility.

The Ecological Basis of Behavioral Flexibility

The costs and benefits of behavioral flexibility will depend, in part, upon the organization of individuals in time and space, and most primates are obligately social. However, even in the Primate Order, noted for the elaboration of social mechanisms (Jones, 2002a; Fleagle, 1999), numerous groups are primitively social (Jones and Agoramoorthy, 2003), failing to display reproduction exclusively by one or a very few individuals. Thus, the advantages of sociality, as Alexander (1974) has pointed out, are often outweighed by its disadvantages, and the changing optima for individuals across time and space may sometimes favor sociality, sometimes not. While primatologists have often assumed that behavioral flexibility is positively and necessarily correlated with sociality, to my knowledge, this assumption has not been empirically supported.

The terms "social" and "sociality" are used in at least two ways in the literature on the evolution of social behavior. One of these is the "Triversian" (Trivers, 1985) definition implying that sociality is any interindividual interaction falling in the categories "selfish," "cooperation," "altruistic," or "spite." The second is a "West-Eberhardian/Alexandrian" definition (West, 1967) whereby the definition of "social" is more or less limited to responses assisting the reproduction of a conspecific. Although variants of the second usage of this term are predominant in the primate literature, in the present volume, an inclusive, Triversian definition will be followed except where noted otherwise, and the relative costs or benefits to inclusive fitness of each of his classes of interaction may have different consequences for the display of behavioral flexibility. Each of these states (selfishness, spite, cooperation, altruism), for example, may be more or less likely to correspond to different degrees of environmental heterogeneity (and, possibly, dispersal costs, generation time of a given population, probabilities of successful reproduction, and/or other factors).

Although group life may partially *buffer* individuals from environmental unpredictability (e.g., finding unpredictable food resources, protection from predators), group life creates an alternate, changing environment in which individuals are required to continuously assess the advantages (benefits) and disadvantages (costs) to survival, reproductive success, and/or phenotype of their potential interactions with conspecifics. Indeed, consistent with McCleery's (1978, p. 381) views, I assume in this volume that "each activity performed by an individual can be thought of as incurring a certain probability of death and a certain probability of successful reproduction." This perspective addresses the problem of the tradeoffs animals make between their investment in behavioral flexibility and their own survival, future reproduction, and/or phenotypic interests. Processes of assessment are not necessarily affected by mechanisms of higher cognition but may be induced by "hard-wired" or associative responses to varying thresholds of environmental stimuli, including stimuli endogenous to the individual (see Parker, 1974). The study of behavioral flexibility is subsumed within the category of differential *effort* invested to optimize fitness (Soler, 2001) and/or benefits to the phenotype, whatever the mechanisms inducing these temporally and energetically limiting processes of allocation. Combined with the views in Chapter 2 and of West *et al.* (2001, 2002) discussed below, this schema for the interpretation of behavioral events, in particular, behavioral flexibility, forms the conceptual architecture of the present work.

The "Patch" View of Behavioral Flexibility

The expression of social behavior and other responses is likely to be a function of environmental heterogeneity (e.g., spatial and/or temporal heterogeneity in climate, predators, mates, food resources, or sleeping sites), expressed as differential degrees of temporal and spatial *patchiness* varying in size, evenness, and quality combined with dispersal costs. Primate units (e.g., genes, individuals, groups, populations, species) may, themselves, be viewed as patches ("habitat patch," "habitat island," "population site," "locality") shaped by heterogeneous regimes. The *patch* view of physical and biotic (including social) environments, initiated by Mac Arthur and Wilson (1967) and Levins (1968; also see Lewontin, 1957; Levin, 1976), has received increased attention in recent years due to researchers' attempts to document the effects of habitat fragmentation and other anthropogenic perturbations upon plant and animal species (Hanski and Gilpin, 1997; Clobert *et al.*, 2001). Despite this renewed emphasis upon spatial dynamics in ecology and population biology, there has been relatively little work in primatology on the causes and consequences of environmental heterogeneity within and between primate species (see Kinzey, 1982; Jones, 1987, 1995a,b,c, 1999a; Fleagle *et al.*, 1999; Ostro *et al.*, 2000).

As the quotation at the beginning of this chapter suggests, organizational units above the individual level are emerging properties of the behavior of individuals. First, principles of ecology indicate that the size and composition of groups change in response to environmental heterogeneity and may have important consequences for the survival and fecundity of organisms (Roughgarden, 1979; Pulliam and Caraco, 1984). Population abundance and structure (including group composition and size) are attributes of resource predictability and quality (Roughgarden, 1979) in combination with dispersal costs (Johnson *et al.*, 2003).

High resource predictability combined with high resource quality, relatively homogeneous spatial dispersion of resources, and resource tracking by the animal population is expected to favor resource defense (e.g., contest competition or territoriality) by individuals or small groups, on average, whereas low resource predictability combined with large distance or high variation in distance between patches may make resources indefensible, yielding large average group size and scramble competition (Pulliam and Caraco, 1984; Emlen and Oring, 1977; Schoener, 1971; Roughgarden, 1979). Since temporal unpredictability of resources may be positively correlated with spatial uncertainty ("patchiness"), foraging in groups may reduce average searching time per individual group member. Thus, environmental predictability and, it is expected, behavioral flexibility, will be inversely correlated with group size (Pulliam and Caraco, 1984; Wittenberger, 1980; Schoener, 1971). The advantages of group life (e.g., increased predator defense, increased efficiency in location of food or mates) should also increase directly with an increase in group size (Pulliam and Caraco, 1984; Wittenberger, 1980; see Johnson *et al.*, 2003), all other things being equal.

Population structure has significant consequences for genes and the individuals who carry them (Hewitt and Butlin, 1997) which may be evident as subdivision into demographic subunits (e.g., aggregations) (≥ 1) or reproductive groups (≥ 2) representing an ecological or evolutionary compromise among those parameters yielding optimal inclusive fitness to individuals (Wilson, 1975; Wittenberger, 1980; Dunbar, 1996; Pulliam and Caraco, 1984). As Wilson (1975) points out, the frequency distribution of group sizes will be a function of those phenomena leading individuals to join and to leave groups combined with the selection pressures on individual responses to these forces (e.g., behavioral flexibility). The parameters determining modal group size, thus, are ultimately expressed as adaptations of individuals to local environments (Wittenberger, 1980; Wilson, 1975; Dunbar, 1996).

In the same local conditions, males and females may adopt different adaptive tactics and strategies due to the energetic and temporal constraints of *anisogamy*, differential investment in gametes between the sexes (Trivers, 1972). Anisogamy has consequences for group size since, all other things being equal, females are expected to adopt those behavioral programs conferring the greatest benefits from the conversion of resources, especially food, into offspring, while the distribution of males is expected to map onto the dispersion of females or their resources in order to optimize fertilization success (Emlen and Oring, 1977; Bradbury and Vehrencamp, 1977; Wrangham, 1980; Wittenberger, 1980; Schoener, 1971; Nunn, 2003). While the availability of energy will ultimately limit group size for populations in density-dependent conditions (Wilson, 1975; Wittenberger, 1980), modal population structure and female social relations are expected to be a function of resource distribution, abundance, and quality in time and space as well as other factors (e.g., predation, male coercion, habitat saturation; Sterck *et al.*, 1997; Wrangham, 1987; van Schaik, 1989). Behavioral flexibility in males, then, is expected to be especially sensitive to the temporal and spatial dispersion (and quality) of females while behavioral flexibility in females is expected to be influenced most particularly by energetic factors, especially the temporal and spatial dispersion (and quality) of food (see Shuster and Wade, 2003). The expected and differential tactics and strategies of females and males permits a preliminary schema (Fig. 1.1) of time and/or energy investment in behaviors along a continuum expected to correspond to likelihood's of damage (after Jones, 1983a) and the differential combination and recombination of behaviors over

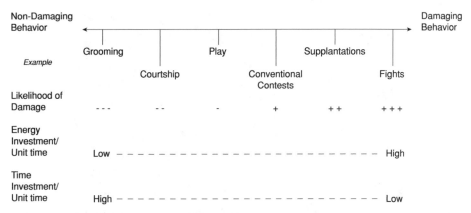

Fig. 1.1. Behavioral continuum reflecting relative time and/or energy investment per unit time from relatively nondamaging to relatively damaging behavior (see Jeanne, 1972; Parker, 1974). Tactics and strategies are hypothesized to differ by sex. [Figure after Jones (1983a, Fig. 3), used with permission].

time and space. Chapters 5 and 6 of this book assess the extent to which these expected profiles may be a function of environmental heterogeneity and the degree to which environmental heterogeneity might favor behavioral flexibility among individuals of each sex.

Environmental Heterogeneity and the Evolution of Behavioral Flexibility in Alouatta palliata

The previous discussion highlights the importance of determining what features of the environment influence the behavioral phenotypes of organisms (West-Eberhard, 1979, 2003; Goldsmith *et al.*, 2003). Following Roughgarden's (1979, Table 13.1, p. 271) schema, for example, mantled howler monkey (*Alouatta palliata*, Fig. 1.2) populations would experience "fine-grained" environmental variation if seasonal rainfall cycles (≈6 months apart) are the primary selective agent since this periodicity is small relative to estimated generation time (T ≈ 6.27 years; Jones 1997b). If the selective agent, say, the periodicity of drought, were the selective agent for the genetic strategy of *A. palliata*, their life history should be subject to "coarse-grained" environmental variation since droughts would probably occur in longer cycles than 6.27 years.

Roughgarden's (1979; also see Gillespie, 1974; Stearns, 1992; Hoffmann *et al.*, 2003) predictions of optimal genetic strategy in each of the four combinations of regimes (where the optimal life history strategy is a function of generation time relative to the scale of temporal and spatial environmental variation) suggests that mantled howler populations experience fine-grained

Fig. 1.2. Adult mantled howler monkeys (*Alouatta palliata*) in Costa Rican tropical dry forest. Photo depicts a consort pair (male, top; female, bottom). Note the vigilant demeanor of male. ©Clara B. Jones.

variation which is best responded to by genetic monomorphism for an intermediate phenotype. This prediction is consistent with what is known of mantled howler genotypes and behavior at Hacienda la Pacífica, Costa Rica (Malmgren, 1979; Jones 1997b, 1999a, 2000). Despite a highly monomorphic genotype, the behavioral flexibility of mantled howlers (and other members of the genus *Alouatta*) is legendary (Crockett and Eisenberg, 1987; Jones, 1980, 1995b, 1999a, 2000; Crockett, 1998; Silver and Marsh, 2003; Bicca-Marques,

2003; Wang and Milton, 2003). Thus, the expression of behavioral flexibility is not dependent upon a highly polymorphic genotype. Indeed, genetic monomorphism may facilitate adaptation to heterogeneous regimes by "buffering" howlers from environmental perturbations (Jones, 1995b). Theoretical treatments of these and related ideas can be found in Emlen (1973), Roughgarden (1979, 1998), and Case (2000).

Conclusions

Wilson (1975, p. 5; also see Odum, 1971, p. 34; Bergman and Siegal, 2003; Stearns, 2003) presents a preliminary schema of factors influencing the expression of behavioral flexibility, including environmental parameters, individual parameters, and population parameters. Each of Wilson's factors displayed (e.g., abiotic, biotic) have been discussed in the current chapter and may exhibit a range of variability (e.g., climate, population density) with the potential to favor or disfavor intraindividual behavioral variation. It is important to note that stochastic parameters may also create conditions of extreme environmental heterogeneity and that an individual's tradeoff of benefits to costs to fitness will be a function of T (Roughgarden, 1979; Emlen, 1983). Wilson's treatment also highlights work by Proulx (1999) and Hoffmann *et al.* (2003), among others (Brooks and McLennan, 2002), who show that phylogenetic factors may bias responses to environmental perturbations and set limits on an individual's potential to respond to them. Some female cercopithecines, for example, have apparently lost the capacity for colonization (see Chapter 2). In subsequent chapters, these factors will be explored further through conceptualizations of the relationship between behavioral flexibility and temporal and spatial environmental heterogeneity. This exercise begins in Chapter 2 with a discussion of one common response by individuals to environmental heterogeneity, dispersal, and an overview of certain theoretical assumptions informing this book.

The Costs and Benefits of 2
Behavioral Flexibility to
Inclusive Fitness: Dispersal as
an Option in Heterogeneous
Regimes

[C]hanges in the conditions of life give a tendency to increased variability; and ... this would be favourable to natural selection.

Darwin (1859, p. 63)

Introduction

The current worldwide biodiversity crisis (Myers *et al.*, 2000; Pimm and Raven, 2000) provides a natural laboratory for the study of behavioral flexibility (Tilman, 1999). Increasing temporal and spatial environmental variability and the effects of anthropogenic factors, in particular, habitat destruction, and subsequent habitat fragmentation and patchiness (Jones, 1999a; Fukuda, 2004) are well documented for numerous primate species (Cowlishaw and Dunbar, 2000; Jones, 1983b, 1995b, 1996b, 1997c, 1999a; Harcourt *et al.*, 2002; Clarke *et al.*, 2002; Fukuda, 2004). This book relies heavily upon theoretical and empirical work on the causes and consequences of biodiversity from the fields of conservation biology and community ecology in an attempt to formulate and suggest tests of ideas appropriate for research at the individual level of analysis. From an evolutionary perspective, dispersal is of fundamental importance since it may counteract the effects of genetic drift by maintaining the connection between subpopulations and populations, preventing isolation.

17

As an alternative behavioral tactic, assessment of the differential costs and benefits of dispersal relative to local conditions highlights patterns of behavioral flexibility with important implications for genotypic and/or phenotypic success. Constraints on dispersal may also facilitate the evolution and/or expression of higher grades of sociality since it has been suggested that some patterns of cooperative and helping behavior occur where habitats are saturated, limiting opportunities for independent reproduction (Emlen, 1994, 1995).

Dispersal from one location to another may occur in response to environmental, especially, temporal, heterogeneity, including local competition, representing one category of flexible behavioral response, and species comprising individuals with limited dispersal capabilities may be at an evolutionary disadvantage if changes occur in the environment that would make movement from one location to another beneficial to survival, inclusive fitness, and/or phenotypic success. Ecological theory predicts a tradeoff between competitive and dispersal abilities which may promote local coexistence of competing species (Amarasekare and Nisbet, 2001). As such, the fitness of specialists is generally greater than the fitness of generalists in the same conditions (Sultan and Spencer, 2002). The present treatment explores the possibility that individual primates experience such tradeoffs, enhancing the benefits of behavioral flexibility and increasing the likelihood that similar genotypes will coexist in the same population. This condition may have particular significance for the coexistence of kin (see Chapter 8) or individuals with similar phenotypes (Chapters 8 and 9). Most important for relatives and nonrelatives is the view that behavioral flexibility may be "functionally adaptive" in temporally and spatially heterogeneous regimes (Travis, 1994).

Dispersal as Flexible Behavior

Figure 2.1 provides one model demonstrating differential (hypothesized) costs and benefits to fitness from behavioral flexibility as a function of differences in home range quality for mantled howler monkeys. This model suggests that decisions to disperse or dispersal imposed by a conspecific (see, for example, Hager, 2003a, b; Jones, 2004; Helms Cahan *et al.*, 2002) occur in response to some threshold of differential benefits and costs to survival, inclusive fitness, and/or phenotypic success. Johnson *et al.* (2003) advance the idea that group formation and maintenance is a combination of resource heterogeneity, abundance, and quality in addition to dispersal costs. West *et al.* (2002) have recently shown that behavioral decisions by individuals, for example, whether or not to disperse, are not only a function of the consequences of the act upon the fitness of one's immediate descendents and other relatives, but also the fitness of others affected by the act, mediated by the

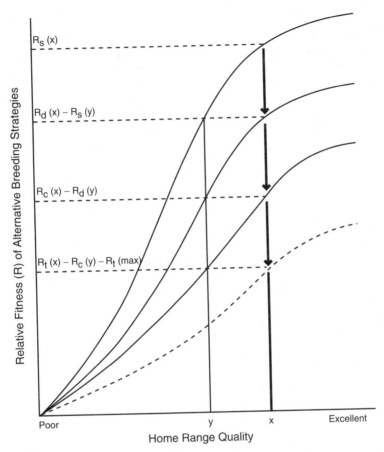

Fig. 2.1. The dispersal threshold model for mantled howler monkeys. R = the fitness of one strategy divided by the fitness of another. The home range quality for a population is determined by ranking home ranges for a population along the abscissa while plotting the relative fitness of group members who adopt different breeding strategies along the ordinate. The values for each alternative breeding strategy depend upon the quality of the home range (x) in the group where each individual lives relative to individuals adopting other strategies in the same conditions (R). Thus, staying and breeding in one's natal group (R_s) yields higher relative fitness than dispersing and breeding in a nonnatal group (R_d) in an environment of a given quality where $R_s(x) > R_d(x) > R_c(x) > R_t(max)$ (the maximum relative fitness the poorest strategy can achieve). The vertical differences between curves (e.g., $[R_d(x) - R_c(x)]$) represent the cost of leaving one's group to adopt a strategy other than breeding in one's natal group. The horizontal distances between curves x and y represent the minimum threshold differences in home range qualities above which it pays an individual to adopt an alternative breeding strategy. The poorer the home range the greater the benefits from alternative breeding strategies and the more likely that unrelated individuals will coexist in groups. (Jones, 1995c; graph based on Emlen, 1994 with permission from Elsevier).

consequences of local competition. The following discussion uses dispersal as a model response to describe West *et al.*'s (2002) treatment and to highlight the potential utility of behavioral flexibility in certain environmental regimes. Patterns of primate dispersal may reveal numerous details about the evolution and maintenance of plasticity in primate populations since theoretical formulations show that where dispersal rates are sufficiently high, behavioral flexibility is favored over specialist tactics and strategies relative to local conditions (Sultan and Spencer, 2002; also see Clark, 1991; Van Tienderen, 1991; Van Tienderen and Koelewijn, 1994).

In order to survive and reproduce, individuals must occupy suitable habitat, and numerous factors determine the likelihood that animals will remain to reproduce on their territories or home ranges. Individuals often disperse as juveniles or adults to new locations, and dispersal rates and distances are expected to determine patch occupation and extinction rates (Wiens, 2001). As pointed out previously, behavioral flexibility is expected to be a stronger effect for overall phenotypic variance for younger than for older individuals, all other things being equal, and, thus, thresholds of response in heterogeneous regimes are likely to vary with age and reproductive value. It is well documented that a wide variety of temporal and spatial modifications of habitat may affect patterns of animal movement, and habitat fragmentation has received particular attention, in part because of researchers' emphasis upon spatial subdivision in metapopulation models (Wiens, 2001; Hanski, 1994; Jones, 1995d). These "source-sink" models show that the frequency of dispersal decreases monotonically with distance from the dispersal source, and numerous studies support the relationship between distance in between fragment and source and the likelihood of a fragment being occupied (Wiens, 2001), factors that could influence and be influenced by an individual's potential for behavioral flexibility. Simple models, however, are often not sufficient to predict the movement patterns of animals because the landscape is, in fact, generally a mosaic of fragments whose components and interrelations may have complex consequences for the behavior of individuals and because individuals in species with different dispersal strategies (e.g., frequent colonizers vs. infrequent colonizers) may respond differently to the landscape "matrix" (Wiens, 2001).

Theory predicts that temporal variation of the environment should select for dispersal (Johnson and Gaines, 1990). Thus, dispersal should be associated with ongoing alterations of habitat such as those occasioned by deforestation, fragmentation, and other anthropogenic effects (Macdonald and Johnson, 2001). By contrast, fragmented habitats demonstrating a static mosaic of spatial heterogeneity should, theoretically, not be associated with increased rates of dispersal (Johnson and Gaines, 1990). This condition may explain researchers' failure to observe increased rates of dispersal in Assamese macaques (*Macaca assamensis*; I.S. Bernstein, personal communication) at abandoned temples overgrown by vegetation in India. Research has shown

(Macdonald and Johnson, 2001), however, that, while habitat fragmentation generating temporal environmental heterogeneity may increase dispersal rates for individuals of species comprising "flexible dispersers" (e.g., mantled howlers of either sex), rates of dispersal may decrease in response to these same conditions by individuals of species comprising "cautious" dispersers (e.g., most female cercopithecines). Flexible dispersers are individuals that disperse readily; whereas cautious dispersers are individuals characterized by reluctance to move away from home ranges or territories.

Case Studies of Behavioral Flexibility

Macaca fuscata

Fukuda (1983, 1988, 2004) observed increased dispersal rates in association with temporal environmental changes by both male and female Japanese macaques (*M. fuscata*: Fig. 2.2). He reported dispersal after a decrease in provisioned food in this species, which typically exhibits female philopatry (Pusey and Packer, 1987). Old, adult females, young males, and individuals of low rank were most likely to emigrate from groups, apparently in response to perturbation of the population's food supply. Fukuda (1988) concluded that the proximate cause of female dispersal was competition for food while that for males was competition for mates, consistent with sexual selection theory (Emlen and Oring, 1977; Strier, 2003). Following the argument of Jones

Fig. 2.2. Adult male (with banana) and female Japanese macaques (*Macaca fuscata*). Females of this species are generally philopatric; however, habitat disturbance has increased incidences of dispersal for both sexes (see Fukuda, 2004). ©Fumio Fukuda.

(1999a; also see Jones, 1995b), Fukuda's (1983, 1988, 2004) results are consistent with the view that dispersal results from a threshold of benefits to costs for individuals and that all individuals (with the possible exception of dependent immatures) are potential dispersers, even those age/sex classes that are generally cautious (e.g., female *Macaca*).

Fukuda (1988) found that group size decreased dramatically consequent to increases in dispersal rates in his study group. He attributed this failure of recruitment to the loss of adult females occasioned by decreased supplies of food. It is possible that female emigration created an unstable group size from whose perturbations the group was unable to recover, increasing extinction risk (Hanski, 2001) and possibly representing one example of the deleterious consequences of behavioral flexibility. Fukuda's (1983, 1988, 2004) studies are perhaps the first empirical documentation of the effects of environmental disturbance on primate group processes (but see Ali, 1981, cited in Pusey and Packer, 1987), including the significance of behavioral flexibility by individuals accommodating to habitat fragmentation.

Alouatta palliata

Although early investigations of mantled howling monkeys (*A. palliata*: Fig. 1.2) reported female philopatry in the species, similar to Old World baboons and macaques (Glander, 1975; Scott *et al.*, 1978), systematic observations of dispersal in mantled howlers demonstrated empirically that bisexual dispersal is the norm (Jones, 1978, 1980; Glander, 1992). This pattern of mobility has subsequently been established for other species of the genus (Crockett, 1984; Brockett *et al.*, 2000a and references therein). Howlers are generally regarded as flexible dispersers (Crockett and Eisenberg, 1987; Jones, 1995b,c; Crockett, 1998) and have been characterized as pioneer species (colonizers) (Eisenberg, 1979; Crockett and Eisenberg, 1987; Jones, 1995b,c; Crockett, 1998), a flexible suite of responses that typically are not expressed by female cercopithecines. Many of the female cercopithecines, apparently, have lost the capacity for, or demonstrate reduced propensity for, colonization in the sense that they are typically philopatric or disperse only into existing groups (e.g., *Papio hamadryas*; see Pusey and Packer, 1987).

A recent report by Clarke *et al.* (2002; also see Clarke and Glander, 2004) showed that deforestation and subsequent habitat fragmentation led to an increase in dispersal rates for female mantled howlers in Costa Rican tropical dry forest, a highly heterogeneous environment for these animals (Jones, 1996b) comprising two habitats (riparian and deciduous; Frankie *et al.*, 1974). Following the argument presented by Jones (1999a), it can be assumed that, in the conditions reported by Clarke and her colleagues, the dispersal threshold was reached as a result of deforestation for an increased proportion of adult females in the study population, but not for adult males or juveniles. Clarke *et al.* (2002) document loss of important feeding trees on the home range of their study group, and perturbation of the food supply may have led to the increased rates of female dispersal.

This interpretation is consistent with Fukuda's (1983, 1988) conclusions for Japanese macaques but fails to identify why dispersal rates failed to increase for adult male and juvenile howlers in Clarke *et al.*'s (2002) study. Because females are expected to be "energy maximizers" and males, "time minimizers" (Schoener, 1971), all other things being equal, it might be speculated that the time budgets of males and the time and/or energy budgets of juveniles were not significantly impacted by the deforestation event reported by Clarke and her colleagues. It would be instructive to use howler groups with analogous characteristics (e.g., group size and composition, home range quality) to investigate rates of dispersal as a function of differential intensities of deforestation in order to determine relative dispersal thresholds of adult females, adult males, and juveniles.

Although the forgoing analysis appears to be straightforward, Macdonald and Johnson (2001) argued that dispersal will be a function of several factors, in particular, the spatial scale and specific characteristics of the environment, as well as life-history parameters of the species. Further, under similar conditions, individuals of the same or different species may not respond in the same manner. These *caveats* highlight the need for intensive empirical work to complement the large body of theoretical literature on the topic of dispersal (Clobert *et al.*, 2001). As Macdonald and Johnson (2001, p. 363) point out, "many animal populations are undergoing declines (or increases) in densities following anthropogenic change and therefore exhibit dynamic dispersal parameters as the perturbed populations experience shifting selective pressures." These same factors may exert selective pressures upon individuals' behavioral responses for accommodation, and possibly, adaptation, to changing regimes. Under these conditions, behavioral flexibility may be favored up to some threshold of response within which costs begin to outweigh benefits for the actor's inclusive fitness.

When to Disperse as an Assay for Demonstrating Behavioral Flexibility

The view of thresholds employed in the present work derives from the insect literature and implies that individuals respond (or, "decide" not to respond—not necessarily consciously) when environmental stimuli reach a level (or a range) of individual sensitivity (see Fewell, 2003, p. 1869; Helms Cahan *et al.*, 2002, pp. 209–210). The threshold view of dispersal holds that individual decisions to disperse occur in response to some tradeoff or calculus of benefits to costs (to inclusive fitness) and that all individuals (with the possible exception of dependent immatures) are potential dispersers (Jones, 1999a). Since individuals are expected to disperse from their resident groups, often their natal groups, to escape the negative consequences of living with kin (e.g., inbreeding or resource competition; Perez-Tomé and Toro, 1982; West et al., 2002; Jones, 1999a; Cole and Wiernasz, 1999), dispersal must be viewed within the framework of social evolution (Ferrière and Le Galliard, 2001; West et al., 2002), even where dispersal yields solitary assemblages (e.g., male orangutans; Rodman and Mitani, 1987; Eisenberg, 1981). Interactions with kin and nonkin are expected to yield differential costs and benefits to fitness leading to decisions (conscious or not) about the value to fitness of staying or leaving. Such decisions will operate regardless of the mode of emigration— whether self-imposed and self-directed (as appears to operate in *M. fuscata*) or imposed by others as is often the case in *Alouatta* spp. (Jones, 1978, 1980, 2004; Crockett, 1984; Glander, 1992; Brockett et al., 2000a; Horwich et al., 2001; Sterck et al., 1997). Hamilton's (1964) rule, $rb - c > 0$, predicts when altruistic behavior toward relatives will be favored, where c is the altruist's cost in fitness, b the fitness benefit to the recipient, and r the coefficient of relatedness. Recent extensions of this rule (West et al., 2002) provide a framework for understanding when individuals will make any behavioral decision, including when individuals will decide to disperse.

West et al. (2002; also see Gandon and Michalakis, 2001; Perez-Tomé and Toro, 1982), discussing "cooperation and competition between relatives," show that it is helpful to consider not only the relationship between actor and recipient, as for Hamilton's rule, but also a behavior's fitness consequences for all individuals affected (e.g., all individuals affected by a decision to disperse). These authors extend Hamilton's rule so that $r_{xy}b - c - r_{xe}d > 0$, "where r_{xy} is the [actor's] relatedness to the beneficiary of its [act] (i.e., the standard r), r_{xe} is the [actor's] relatedness to the individuals who suffer the increased competition from the beneficiary ..., and d is the general decrement in fitness associated with the ... act" (West et al., 2002, p. 73). r_{xe} will be associated with local (single patch) competition, increasing as within-patch competition intensifies, a condition that may arise where individuals remain in their local territories, home ranges, or neighborhoods. Assuming here that the act in question is dispersal by an individual group member, West et al.'s new equation will be equal to the original Hamilton's rule when $r_{xe} = 0$ (the disperser is unrelated to his/her competitor's competitors) or when $d = 0$ (competition is not increased) (West et al., 2002). As r_{xe} or d increase, the benefits of dispersal from kin selection decrease (West et al., 2002). Importantly, a consequence of these increments, inferred from West et al. (2002), is that the disperser will be *effectively* more closely related to the disperser's competitor's competitors, and the *effective* relatedness between relatives will be decreased, decreasing, for example, the benefits of dispersal.

The inference from West et al.'s (2002) analysis shows that, if $r_{xy} = r_{xe}$, the individual will not "decide" to emigrate since the costs to fitness from dispersal would outweigh the benefits because any potential gains to the beneficiary of the act of dispersal would be equaled or outweighed by costs to the fitness of the beneficiary's competitors. Under these conditions and for this response, behavioral flexibility would not be beneficial to the individual. Additional theoretical work is required to demonstrate the precise ways in which r_{xy} and r_{xe} operate in temporally and spatially variable regimes such as fragmented and other disturbed environs, and as West *et al.* (2002) make clear, measuring the parameters in their equation will be a daunting task, particularly for long-lived species such as primates and many other mammals. Nonetheless, knowledge of r_{xy} and r_{xe} in relation to local competition may explain variation in dispersal rates and other responses by individuals within and between populations and habitats. Of particular interest is the possibility that behavioral flexibility, enhanced by emotional and cognitive processes, permit individuals to assess the relative costs and benefits to direct descendants, other kin, and kin's competitors in relation to environmental regime. Langurs might be a model taxon for investigations of the causes and consequences of these complex decisions since these monkeys demonstrate marked variability in dispersal strategies within and between species (Sterck, 1998, 1999), and their sociobiology is relatively well known. Studies, including experiments, comparing noncooperatively breeding and cooperatively breeding callitrichids would also provide important information on the costs and benefits of, as well as the constraints on dispersal that may have general import (e.g., Abbott, 1998).

Toward a New Perspective on Behavioral Flexibility

West *et al.*'s (2002) treatment leads to the conclusion that costs drive decisions, including decisions to disperse, in particular, costs of assisting the reproduction of conspecifics, especially kin (the "West-Eberhardian/Alexandrian" view of social behavior). These authors' formulations provide a general framework for understanding why individuals act as they do in particular regimes, including decisions to disperse or not to disperse. West *et al.*'s (2002) formulations demonstrate that inclusive fitness depends critically upon local conditions (including genotype frequencies in a population and population density; see Wauters *et al.*, 1994; Treves and Chapman, 1996). West *et al.*'s work is consistent with the view that behavioral flexibility reflects the competitive abilities of individuals responding to heterogeneity (levels of competition) in local conditions and suggests that dispersal, like other types of behavior, is driven by self-interest in relation to the differential costs and benefits of assisting others' reproduction *except, perhaps, where dispersal is manipulated or imposed by conspecifics, including kin.*

As Macdonald and Johnson (2001) point out, however, habitat disturbance (*or, it might be added, any environmental perturbation above a certain threshold*) may create new selection pressures upon individuals, altering patterns of movement (*or, other behavioral response*) within and between populations in novel ways. With the possible exception of Pope's (1992) results for *A. seniculus*, the red howler monkey, no studies of the evolution of dispersal rates or effective rates of gene flow exist for primates, and these data for other animals are "practically nonexistent" (Hanski, 2001). Hanski's (2001, p. 297) review documents that "a high dispersal rate is a prerequisite for a high rate of population establishment and, thereby, for long-term metapopulation survival" and that "dispersal rate is likely to increase when the environment becomes increasingly fragmented and the rate of extinction becomes increased." The responses that animals make when human intervention is not present will significantly influence, if not determine, the persistence of populations in nature as anthropogenic perturbations increase, and species whose individuals exhibit a greater degree of behavioral flexibility are less likely to be "extinction prone," all other things being equal (Jones, 1995b, 1997c). As Sultan and Spencer (2002, p. 280) state, "In general, higher migration rates favor plasticity...." Thus, it will be important to investigate rates of dispersal and gene flow in primates for a thorough understanding of behavioral flexibility.

High rates of dispersal are found in species inhabiting naturally patchy habitats (Hanski, 2001; Pope, 1992), and female dispersal rates appear to vary in response to variations in the physical and biotic regime such as changing likelihoods of infanticide in time and space (Sterck, 1997). Not all species, however, are expected to demonstrate the same flexibility in the face of temporal environmental heterogeneity. As Macdonald and Johnson (2001, p. 367) point out, "the role of dispersal in population dynamics may vary substantially between species," a research program worthy of investigation among primates.

Primate species with a restricted habitat range (β-rarity) may be "extinction prone" (Jones, 1997c; Harcourt *et al.*, 2002; Chivers, 1991). Jones found that, although primates may be differentially sensitive to moisture gradients and the subspecies or races of certain species may show notable habitat specificity (e.g., *Colobus badius*), most species sustain populations in several habitat types, exhibiting a broad tolerance for habitat expansion. This analysis revealed that, while most primates demonstrate broad habitat specificity (low β-rarity), frugivory in combination with the ingestion of other plant tissues appeared to be related to β-rarity and, thereby, vulnerable to extinction. Based upon Jones' (1997c) study, Table 2.1 lists those species that might be expected to demonstrate a significant degree of behavioral flexibility (generalist or polyspecialist strategies) because of their exposure to regimes with extreme temporal and spatial heterogeneity.

Future research should investigate whether individuals of primate species demonstrating restricted habitat specificity are likely to be cautious dispersers, and thus, individuals with a lesser likelihood of behavioral flexibility, while individuals of species demonstrating broad habitat specificity, are likely to be

Table 2.1. A List of Primate Species ("common species") Demonstrating Local Population Densities that are Somewhere Large, Habitat Specificity that is Broad, and Geographic Distribution that is Wide (N = 18; after Jones, 1997c). Number of subspecies or races in parenthesis (after Groves, 2001; Wolfheim, 1983).

Lorisidae
    Galago crassicaudatus (10–11)
    G. demidovii (7)
    G. senegalensis (9)
Callitrichidae
    Cebuella pygmaea (2)
Cebidae
    Alouatta seniculus (4)
    Aotus trivirgatus (8–9)
    Cebus apella (11)
    Saimiri sciureus (8)
Cercopithecidae
    Cercopithecus aethiops (21)
    C. ascanius (5)
    Colobus badius (20)
    C. guereza (9)
    Macaca fascicularis (21)
    M. mulatta (4)
    M. nemestrina (4)
    Papio anubis (4–7)
    P. cynocephalus (3–4)
    Semnopithecus entellus (15)

NB: In addition to human beings (*Homo sapiens*), these species might be most likely among the primates to exhibit behavioral flexibility. Other candidates for the elaboration of behavioral plasticity are species experiencing some combination of these three factors (see Jones, 1997c).

flexible dispersers, and, thus, individuals with a greater likelihood of behavioral flexibility. Studies of these and related topics would permit the identification of particularly vulnerable species and might justify interventions [e.g., translocation ("artificial dispersal"); Jones, 1999b] that could mitigate or delay the most deleterious consequences of anthropogenic perturbations. Additional research is needed to investigate possible relationships between local competition and behavioral flexibility, in particular, the need to test alternative hypotheses for the evolution and maintenance of plasticity in primates. Similar to recent treatments in conservation biology (see Tilman, 1999; Hubbell *et al.*, 1999), likelihoods of behavioral flexibility in primates might be informed by a better understanding of "recruitment limitation" as a function of environmental heterogeneity. Most important, perhaps, the next section aims to show that tactics and strategies of behavioral flexibility are mechanisms for the management of local competition as the individual attempts to optimize his or her inclusive fitness. For some individuals in some conditions, behavioral flexibility can reduce competition or retard its escalation.

Why are There so Many Kinds of Behaviors?

Hutchinson's (1959) paper, "Homage to Santa Rosalia *or* why are there so many kinds of animals?" was a landmark attempt to explain species diversity, a topic still poorly understood (Tilman, 1999). This and other seminal works (e.g., Root, 1967) addressed a species' niche exploitation patterns as adaptations to heterogeneous regimes. Analogous to studies at the level of species, subspecies, and populations, an individual's behavioral repertoire may be viewed as a set of tactics and strategies designed for niche exploitation to optimize fitness. Just as Root (1967, p. 346) proposed the term *guild* for "groups of species that exploit the same class of environmental resources in a similar way," the term *phenogroup* can be applied to individuals exhibiting similar patterns of behavioral exploitation with similar environmental consequences. Assortment may occur on the basis of similarity (e.g., kinship, appearance, response patterns; Dickinson and Koenig, 2003; Parr and de Waal, 1999; Buss, 1984; Sinervo and Clobert, 2003; Baglione *et al.*, 2003, see Chapter 4; positive phenogroup assortment such as some "trait groups", see Wilson, 1980). Assortment may also occur on the basis of difference (negative phenogroup assortment), and it is expected that both positive and negative phenogroup assortment will be expressed relative to an individual's condition-dependent benefits and costs to genotype and/or phenotype.

Phenogroups may be exposed to similarly varying frequencies, durations, rates, and/or intensities of competition for limiting resources, including food, mates, and/or space, and, just as niche segregation within a guild differs in relation to differences in efficiency in patterns of exploitation of guild members, members of phenogroups may be segregated by differences in their efficiencies of exploitation patterns, a condition expected to favor behavioral flexibility to decrease niche overlap in competitive, temporally and spatially heterogeneous regimes. Chapters 3 and 4 address some of the relatively invariant features of primates that might structure phenogroups such as social parasitism and the capacity for "categorization" as well as other responses likely influenced strongly by morphology, age, gender, and size. The problem of why there are so many kinds of behaviors is a problem of coexistence in time and space of exploitation patterns varying within and between individuals. The diversity of behaviors found among primates (e.g., Boesch *et al.*, 2002) in part may be explained, then, by varying intensities of interindividual competition for limiting resources and resulting pressures to reduce competition by the expression of relatively nonoverlapping patterns of resource exploitation.

Conclusions

Future studies of behavioral flexibility in primates need to investigate potential causes of response diversity other than or in addition to changes in temporal and spatial heterogeneity (e.g., nonequilibrium conditions, interactions among species at a variety of trophic levels, and "recruitment limitation"; Tilman, 1999). In addition, it will be important to study instances in which behavioral flexibility is expected but does not occur. For example, it might be expected that cooperatively breeding callitrichids would exhibit reconciliation because repair of relationships should be of great benefit to these highly communal individuals. However, reconciliation is rare or absent in most of these species (Schaffner and Caine, 2000). Research on this and other clear cases that violate expectations about when behavioral flexibility should be found would enhance our understanding of constraints on behavioral flexibility as well as what are probably the many responses compensating for behavioral flexibility. Schaffner and Caine (2000), for example, propose that because cooperatively breeding callitrichids are very communal and display reproductive suppression, reconciliation is not required as a response to resolve conflict or to mend relationships.

In addition to research into the presence or absence of behavioral flexibility across taxa, detailed studies are required on the interaction between competition and dispersal in order to refine the formulations (after West *et al.*, 2002) presented in this chapter. This body of work would test the idea that a tradeoff between competition and dispersal may favor behavioral plasticity to promote the coexistence of individuals with similar genotypes and/or phenotypes locally (Amarasekare and Nisbet, 2001). Finally, studies of primate dispersal rates are needed in order to evaluate whether high rates of dispersal decrease the likelihood that similar individuals will coexist locally and, thus, decrease selection for behavioral flexibility (Amarasekare and Nisbet, 2001).

Primate Signatures and Behavioral Flexibility in Heterogeneous Regimes

3

Organisms capable of extensive modification are termed plastic; and this plasticity may be subject to selection.

Baldwin (1902, p. 94)

Introduction

Scientists are trained to generalize, and many primatologists have attempted to characterize those traits diagnostic of the Primate Order (Vaughan, 1978; Eisenberg, 1966, 1981; Fleagle, 1999; Jones, 2001; Kappeler *et al.*, 2003; Lee and Kappeler, 2003). Of course, the traits that one considers significant will often depend upon one's questions, and the goals and assumptions of research have not been the same in all of these research programs. Thus, although large brain to body ratio and a capacity for learning, in particular, social learning, and cognition are mentioned by many investigators as signatures of primates (e.g., Mazur, 2002; Fragaszy and Perry, 2003a), few of the other traits appear in all schemas (e.g., specializations in dentition or cranial anatomy). This chapter will propose several traits displayed by primates that appear to this author to have received little attention as possible primate signatures and which may facilitate and/or represent behaviorally flexible responses associated with the success of primates in temporally and spatially heterogeneous regimes. After an initial overview of learning mechanisms (Mazur, 2002; Jones, 2002a; Silver and Marsh, 2003), I will discuss time and energy allocation strategies (Brockmann, 2001; Jones and Agoramoorthy, 2003), the elaboration of displays of persistence (e.g., queuing, Alberts *et al.*,

2003; olfactory seduction, see Jones, 2003a) and persuasion (Grammer, 1989; Jones, 1996a), social parasitism and mechanisms of "phenotypic manipulation" (Jones, 1986, 1996a), and the phenomenon of individuality (Bolnick *et al.*, 2003; Jones and Agoramoorthy, 2003)—a primate signature that may dampen the evolution and/or expression of behavioral flexibility and "true" sociality (e.g., eusociality) by constraining the extent to which individuals are likely to inhibit their selfish impulses. If the proposed behavioral features are signatures of primates (and, possibly, other social mammals), then they should be amenable to comparative and other analyses (Crespi and Choe, 1997; Reeve, 2001).

Learning, Environmental Heterogeneity, and Behavioral Flexibility

In the social sciences, use of the term "plasticity" extends at least to the writings of G. Stanley Hall and William James in the late 1800s. Hall was influenced both by the physiological psychology then current at Harvard, by Wundt's laboratory in Germany, and by the evolutionary theory in the form of his "theory of recapitulation." James was also concerned with the adaptive significance of behavior and with individual differences and became identified with the American school of "functionalism" which arose in reaction to the structuralism of Wundt and his followers (e.g., Titchener, Washburn), both in the United States and abroad. The concern for interindividual and intraindividual variation and for adaptation to local conditions links these early researchers and their associates with several traditional subjects of study in psychology, such as individual differences, learning, and development, in addition to physiology, including sensation and perception and, notably, Darwinian theory (Darwin, 1859, 1965). These areas of study continue to be of interest to students of behavioral flexibility relative to features of the environment (Reader and Laland, 2003; Fragaszy and Perry, 2003a; Jones, 2002a,b; Sherman and Visscher, 2002; Deecke *et al.*, 2002; Grieco *et al.*, 2002; Kameda and Nakanishi, 2002; Kudo and Dunbar, 2001; Heckhausen and Singer, 2001; Wright, 2000; Hollis *et al.*, 1997).

Recent research on behavioral flexibility and diversity in the social sciences has emphasized *social learning* (learning by imitation, observation, or modeling), in particular, as this mechanism fosters cultural and traditional responses (Fragaszy and Perry, 2003a; van Schaik *et al.*, 2003; Aureli and Whiten, 2003; Boesch *et al.*, 2002; Whiten *et al.*, 1999; see Silver and Marsh, 2003 on foraging strategies in *A. pigra*). Perry (2003, pp. 429–430) reports that gregariousness and interindividual tolerance promote social learning, and that social learning is most likely to be found in regimes "with intermediate rates of environmental change" (*relative to* T). As discussed in Chapter 8, these factors correspond to the conditions expected to favor the formation of multimale–multifemale and/or cooperatively-breeding assemblages, the highest grades of sociality within the Primate Order, not all of whose representatives demonstrate a noteworthy degree of behavioral flexibility (Box, 1999; see Chapters 2 and 8). Furthermore, extensive behavioral flexibility has been documented for many primate species with dispersed, monogamous, or polygynous sociosexual organizations (African lorises, Hager and Welker, 2001; lemurs, Schulke, 2001; Andres *et al.*, 2003; titi monkeys, Mayeaux *et al.*, 2002; see Chapter 8). Related to this discussion, it will be important to investigate responses other than overt behavioral ones (e.g., genetic, physiological) that may be employed by primates in heterogeneous regimes (e.g., Watson *et al.*, 2003).

While the generalization that behavioral flexibility correlates with social complexity indicates that cultural and traditional action patterns are responsive to local conditions and may be adaptive, their expression may be too labile for their utilization as phylogenetic markers in comparative studies, as advanced by Fragaszy and Perry (2003b). Nonetheless, the papers in Fragaszy and Perry (2003a) strongly suggest that cultural and traditional responses facilitate the optimization of individual fitness in temporally and spatially heterogeneous regimes and that general principles of social learning are amenable to empirical investigation. Like the signatures discussed below, learned responses are expected to be, on average, "effective" in the sense that they facilitate individual survival and reproductive success (Tolkamp *et al.*, 2002), all other things being equal (e.g., in the absence of social parasitism).

Fitness as a "Fixed Budget" of Time and Energy Generating Signatures of Primate Behavior: The Temporal Component

Tolkamp *et al.* (2002; also see Martin *et al.*, 2002; Dunbar, 1992) develop ideas about the optimization of short-term animal behavior in relation to the currency of time, and, where discriminable asymmetries exist between interactants, individuals with certain traits (e.g., subordinates) might exert a significant degree of control over the timing and duration of these events (Johnstone and Bshary, 2002; Jones, 1997a, 2002a). Deriving their formulations from aging theory whereby oxidative metabolism is limiting ("metabolic time"), Tolkamp *et al.* (2002) view behaviors as response sets designed to maximize benefits per unit of "metabolic time" expended. For purposes of the present discussion concerning the relationship between behavioral plasticity in primates and environmental heterogeneity, one aspect of Tolkamp *et al.*'s (2002) analysis is of particular interest, the idea that since time may be defined as a limiting resource, time like energy, then, becomes a resource for which individuals compete.

Making decisions in the short term to influence lifetime reproductive success (Tolkamp *et al.*, 2002; also see Barrett and Henzi, 2002), individuals may behave so as to deplete others' store of time and to minimize error, unpredictability, uncertainty, and risk in their own allocation of time. Examples of the former condition might involve a wide range of habits designed to "waste" or to expend another's time budget [e.g., grooming, Silk *et al.*, 2003; Dunbar, 1997; various patterns of communication (e.g., redundancies in speech or other responses, halting or slow patterns of auditory or other communication, pauses, use of interjections such as, "Ah . . . ." and the like, see, for example, Balistreri, 2003); obnoxious phenotypes, Jones, 1985b; Kowalski, 2003; Balistreri, 2003]. These conditions are expected to obtain only *where it is in the actor's genotypic and/or phenotypic interest to do so* (Vickery *et al.*, 2003) or where individuals have no other behavioral options, similar to states favoring fighting with an opponent to the point of death (West-Eberhard, 1979).

Expending time and energy to induce others to utilize and "stress" their limited reservoir of time resources is not *spiteful* "if it harms others to an extent that is substantially greater than its expected cost" (Vickery *et al.*, 2003, p. 415; Foster *et al.*, 2001) or where it functions as pseudoaltruism in the service of phenotype matching. If the responses in question are harmful to the actor as well as the recipient, however, costs might be mitigated where recipients or onlookers of spiteful behavior interpret these responses as a handicap or as condition-dependent signals of quality, potentially benefiting the actor. Such conditions might obtain in straightforward interactions between conspecifics. Spiteful responses might also be components of compound signals or displays which optimize the actor's inclusive fitness or phenotypic success, on average, or might be expressed if spite is likely, above some threshold, to initiate or lead in the future to compensate beneficial outcomes by the recipient or a third party.

It might be expected that spite has evolved among many social primates, including humans (W.L. Vickery, personal communication) because of the high frequency, duration, rate, and/or intensity of interactions in these species, conditions with the potential to increase the uncertainty and error (heterogeneity) of signals, response sets favoring mixed strategies (Plaistow *et al.*, 2004; see Chapter 4), and behavioral flexibility. Such responses might be most common where likelihoods of escalation or other damaging behavior are high, often associated with "patchy" conditions in which limiting resources (e.g., food, mates, sleeping sites) are clumped in time and/or space (see Fig. 1.1). Primates that are obligately social (e.g., most female primates) might be especially likely to engage in interindividual competition for time since this strategy of agonism is less likely to escalate to damaging fights (Maynard Smith and Price, 1973; Parker, 1974; see Chapter 8), a consequence that would be energetically costly and disruptive of social relations.

Persistence and Persuasion as Features of Primate Time Budgets

Ultimately, organisms are limited by the currencies of time and energy in their efforts to survive and reproduce. One characteristic of primates that may represent a signature of the order, including males in mating contexts, is *persistence*, a strategy that may be costly in time since an individual waits for an opportunity to act (e.g., to feed, to mate; Smuts, 1985, 1987a; Clarke, 1990; Jones, 1997a; Alberts *et al.*, 2003). Alberts *et al.* (2003; also see Hager, 2003a) have recently described "queuing" by male savannah baboons (*Papio cynocephalus*) whereby males waited for opportunities to mate. Effects of queuing were particularly evident over the short term when copulation success was assessed in relation to dominance rank, as might be predicted by Tolkamp *et al.*'s (2002) model if it is assumed that higher ranking males have a larger time budget and that males utilize queuing to minimize costs of competing directly (e.g., with expensive displays or by fighting) for access to females (see Fig. 1.1).

Interestingly, Alberts *et al.* (2003) show that the queuing system broke down as the number of males in a group increased ("queue-jumping"), suggesting that this facultative response is sensitive to local conditions such as interaction rates. Queuing as a mechanism of persistence, then, is most likely to be associated with stable conditions, and queue-jumping is the behavioral response most likely to be associated with higher levels of environmental (within group) heterogeneity (Alberts *et al.*, 2003, p. 833; Hager, 2003a). Importantly, Alberts *et al.* (2003) show that their results have important implications for the dispersal strategies of males and that these effects are likely to be complex. Nonetheless, these authors predict that young males will favor small groups and the benefits of a queuing strategy. Research is required to determine how widespread this strategy of dispersal is among young primate males.

Once persistence has become "fixed" in a population (by selection and/or by culture or tradition), the opportunity arises for "functionally adaptive" tactics and strategies of *persuasion* to arise for the optimization of fitness by individuals (Barrett *et al.*, 1999, 2000; Jones, 1997a, 2002a,b,c, 2003a; Miller, 2000). Discussing sexual conflict, Brown *et al.* (1997; also see Clutton-Brock, 1998) argue that persuasion is one strategy that may be employed to manipulate an interaction where conflicts occur between interacting individuals with different fitness optima. Persuasion may lead to one individual offering a fitness incentive to another, say, in the form of an increased share of group (reproductive) output (Hager, 2003a). However, incentives might be any activity, response, or event that would increase the time and/or energy budgets of an individual or serve another's interests in terms of survival and/or reproduction (even indirectly through phenogroup assortment). Numerous reports on primates suggest that the responses displayed by individuals in certain of these species are social in the sense defined by Brown *et al.* (1997: i.e., persuasive interindividual interactions; for example, Silk, 1987; Walters and Seyfarth, 1987; de Waal, 1989, 1990; Abbott, 1993; Thierry, 2000).

Thierry (2000) concluded that the pattern of results observed in his study of conflict management strategies in macaques was consistent with van Schaik's (1989) socioecological model implicating predation pressure as the determining selective pressure favoring "despotic" or "egalitarian" populations. In general, despotic conflict management was associated with high risks of predation (high temporal and spatial heterogeneity of predation pressure), and egalitarian conflict management was associated with low risks of predation (low temporal and spatial heterogeneity of predation pressure). Further research is required to test Thierry's (2000) inferences and to document the relationship between intraindividual variations in patterns of conflict management and environmental heterogeneity in macaques.

It would be expected, however, that individuals in egalitarian species would be most likely to employ persuasive strategies in their attempts to resolve and to manage conflict. Persuasion, then, is expected to be a signature of individuals of egalitarian species (e.g., *Brachyteles*; *Macaca maura*) and of individuals in conditions where providing incentives to subordinates is beneficial to fitness (see Jones, 1997a). Hager's (2003a) review shows that these conditions are most likely to be found in multimale–multifemale groups, assemblages that have formed in response to the benefits males derive from close association. Further research is required to document an association between male–male persuasion and multimale–multifemale groups and female–female persuasion and female groups. Nonetheless, it is expected that heterogeneous conditions favoring group formation (Johnson *et al.*, 2002; Jones, 2004) would favor a broad range of flexible persuasive responses (e.g., communication systems, alternative reproductive tactics and strategies).

Social Parasitism as a Signature of Primates in Heterogeneous Regimes

The investigation of *parasitism* generally entails interspecific invasion of hosts by viral, bacterial, protozoan, or other small organisms which utilize the host's tissues, blood, or other products for survival and/or reproduction (see Altizer *et al.*, 2003; McRae, 1997). Usually, the survival and reproduction of the parasite depend upon the survival of the host; thus, parasitism is differentiated from predation. The concept of *intraspecific social parasitism* initially arose in the insect literature to describe the potential for members of the same species to evolve exploitative mechanisms with relatively permanent effects upon victims (hosts). These mechanisms are presumed to be employed to optimize fitness of the social parasite in response to interindividual (genetic) conflicts of interest, leading the social parasite to exploit the responses of its host(s). Intraspecific social parasitism is expected to be beneficial to the social parasite in heterogeneous regimes as a condition-dependent, reversible tactic to optimize likelihoods of survival and reproduction.

The present treatment of intraspecific social parasitism (ISP) as a (facultative) primate *signature* relies heavily upon the literature on social parasitism in insects (e.g., West-Eberhard, 1986; Wcislo, 1987; Hölldobler and Wilson, 1990; Cichon, 1996) as well as analogies from the literature on classical or nonsocial parasitism. For example, like the study of herbivory and predation, parasitism—and, by extension, social parasitism—is defined primarily in terms of its costs (Bronstein, 2001, 2003). Importantly, Bronstein's (2003) research demonstrates ways in which costs to reproductive success from associations such as social parasitism can be quantified, empirical data that are needed for primates and other social vertebrates. Further, in the literature on nonsocial parasitism, it has been shown that parasites and nonparasites may coexist within the same population (Maruyama and Seno, 1999; Wahl, 2002; Haccou et al., 2003; Poulin, 2003), and the present treatment assumes, likewise, that some individuals in a population may express social parasitism while others may not. Populations, then, may be polymorphic for social parasitism. On the other hand, since most adult primates will exhibit a capacity for totipotency, they are expected to facultatively express social parasitism—and respond to it—when it is in their interests to do so. The present treatment of social parasitism is particularly concerned with the hypothesized relationship between social parasitism "as an engine of diversity" (Summers et al., 2003).

Based upon studies on hymenopterans, Savolainen and Vepsäläinen (2003) suggested that polygyny is a prerequisite for ISP and that social parasites are often related ("Emery's rule"), conditions that may pertain, as well, to primates. Related to Savolainen and Vepsäläinen's recent work, Crespi and Choe (1997a,b) provided ideas about ways in which nonsocial parasitism and reproductive skew may be associated, treatments applicable as well to ISP. Studies on primates should investigate the possible association between ISP and different grades of sociality (e.g., cooperative breeding) in the Order.

In heterogeneous regimes, ISP may increase the advantages of behavioral flexibility since increased variability of responses may diversify response repertoires, buffering individuals against the potentially deleterious effects of social parasitism and other forms of exploitation. Poulin and Thomas (1999) review the literature on nonsocial parasitism describing host traits modified by parasites and their evolutionary consequences, including behavioral effects. These research findings, combined with other reports (e.g., Summers et al., 2003), should be explored for an understanding, by analogy, of the causes and consequences of ISP in primates and other social vertebrates and their potential benefits under conditions of environmental uncertainty. Displaying behavioral flexibility by the expression of intraspecific social parasitism, for example, may reduce for potential competitors the cost of food, mates, or other limiting resources since the host is, in effect, an alternate resource for the social parasite.

Theoretical models of nonsocial parasitism, then, may also apply to ISP. Wahl's (2002) model showed, for example, that social parasitism favors evolution of unfair division of labor since social parasites do not contribute to group labor or reproduction. In Jones's study of Costa Rican mantled howler monkeys (Jones, 1980, 1985a), the second-ranking male in the adult hierarchy rarely participated in group defense with other males (C.B. Jones, 2000, unpublished data). This male was expelled from his group as the result of a coalition between the dominant male and a new, young male (Chapter 6). Failure to cooperate with other males (ISP) may reflect a benefit (energy and/or time savings) while expulsion may reflect a cost (expulsion) of ISP, and individuals are expected to assess these potential tradeoffs (consciously or otherwise) before committing to a response, a calculus entailing varying amounts of uncertainty and risk (e.g., of escalation or expulsion).

Jones (1997a) reported social parasitism by cycling female mantled howler monkeys (*A. palliata*; also see Linklater and Cameron, 2000 for *Equus caballus*, the feral horse; Clutton-Brock, 2002). Females of this species sometimes exploit males by leading them to feeding sites with displays of sexual receptivity (the "rear-present" posture). Once at a feeding site, females may or may not feed and copulate with males (see Fig. 1.2). Social parasitism, then, occurs when one individual exploits another's resources, including a conspecific's cooperative relationships, and it is expected that the social parasite will receive relatively long-term genotypic and/or phenotypic benefits at another's expense, usually without killing it.

Parasitoids, on the other hand, may kill their host (e.g., some cases of homicide by stepfathers; Daly and Wilson 1988), showing that parasitic and parasitoidal relationships can vary in virulence. One might speculate, then, that *social parasitoids* are ISPs whose behavior leads to the failure to reproduce, or results in the death of the host or victim. In the *A. palliata* example, a female exploits a male's ability to defend a feeding site. After feeding, a female may or may not copulate with the parasitized male. Female responses, then, may negatively impact a male's reproductive success and, possibly, lead him to adopt potentially damaging (i.e., riskier) behaviors (e.g., fighting) to obtain access to fertilizable females (Jones, 1995a, 2002b). Females may benefit from social parasitism of males wherever it is difficult for a female to assess a male's response.

Social parasitism represents one potential cost of sociality whereby individuals may be exposed to increased exploitation from others as a result of proximity. As suggested above for primates, evidence from parasitic birds suggests that social parasitism may be expressed most likely in temporally and spatially heterogeneous regimes (Jamieson *et al.*, 2000). Social parasitism, then, concerns relationships of dependency which may vary from associations in which the parasite cannot survive without the labor of its host (e.g., some parent-offspring relationships; Galef, 1981; Nicolson, 1987; Clarke, 1990; Trivers, 1974) to relationships in which parasites effectively displace hosts (e.g., some group takeovers: patas and forest guenons, Dixson, 1998, p. 64; gorilla, gelada, hamadryas, Dixson, 1998, p. 66, Fig. 4.12). Social parasitism (e.g., phenotypic manipulation, pseudocooperation, pseudoaltruism) may be a signature of primates because most species of the order are obligately social, making individuals highly vulnerable to the potential for manipulation and exploitation occasioned by the costs of sociality (de Waal, 2000; also see Combes, 1995; Crespi and Choe, 1997a,b).

Phenotypic Manipulation as a Form of Social Parasitism

One mechanism of intraspecific social parasitism that may be particularly elaborated among primates for interindividual exploitation in heterogeneous regimes is "phenotypic manipulation" or "behavioral alteration" (Jones, 1986; Cialdini, 2000; Poulin, 2003; Trivers, 1974) whereby a conspecific exploits the phenotype of another with relatively enduring benefits to the victimizer, presumably at the ongoing expense of the victim. Here, we are primarily concerned with the exploitation of behavioral responses and its implications for behavioral flexibility. Phenotypic manipulation may generally lead to maladaptive responses by the victim (e.g., self-destructive behavior, homosexuality, various exogenously induced psychopathologies). These responses are expected to optimize the social parasite's or a third party's, rather than the victim's, client's, or host's, individual (genetic) self interests. While some authors have discussed "the evolution of maladaptation" as a potentially adaptive response (Crespi, 2000), phenotypic manipulation implies that it is in the interests of individual A to exploit the phenotype of individual B for potential long-term gain. It is expected that individual B would attempt to resist such manipulation and that a coevolutionary race might evolve between exploiters and those most likely to be exploited (e.g., between relatives, between adults and young, between dominants and subordinates).

For example, where homosexuality or other nonreproductive alternative sexual phenotypes result from phenotypic manipulation, manipulated individuals might counteract the negative (genetic and/or phenotypic) consequences of these phenotypes by investing in kin and/or their offspring or by adopting young whose phenotypes might, in turn, be manipulated to benefit the adoptive parent(s), their relatives, and/or matching phenotypes of unrelated individuals. On the other hand, while genotypic (e.g., via inclusive fitness) and/or phenotypic (e.g., via "green beard" effects) benefits may accrue to exploited individuals adopting counterstrategies to a social parasite, observations of same sex partner preference in adolescent primates (see Vasey, 1995) suggests that the expression of these behaviors may simply be driven by endocrine maturation and may not be adaptive. Nonetheless, as

Fig. 3.1. The potential exists for individuals to exploit one another where they live in groups and interact frequently. This female stump-tailed macaque (*Macaca arctoides*) may be grooming the infant's mother in an attempt to influence mother and/or her offspring. ©Mike Seres.

discussed above, it is important to investigate the genotypic and phenotypic consequences of behaviors for the actor's fitness and for the fitness of all recipients of the act(s).

The long developmental periods of most primates expose their phenotypes to manipulation by conspecifics, including parents, caretakers, and socially dominant peers (Fig. 3.1). Indeed, these long developmental periods may have evolved in part to serve parents', especially mothers', fitness optima by affording her opportunities to shape offspring phenotypes to serve her selfish genetic interests (e.g., "maternal effects"; Mousseau and Fox, 1998a,b). Recent research with insects and birds suggests that female mate preferences are sexually selected and driven by her expected costs and benefits in particular environmental regimes (Kotiaho *et al.*, 2003; Qvarnstrom *et al.*, 2000), outcomes that might partially determine a female's ability to selfishly manipulate her offsprings' phenotypes.

Table 3.1 presents possible examples of social parasitism and phenotypic manipulation in primates. It may be possible to investigate the presence or absence of mechanisms of social parasitism with operant techniques since intraspecific social parasitism (ISP) should be a function of learned or other nongenetic processes (e.g., behavioral suppression by olfactory substances).

Table 3.1. Possible Examples of Social Parasitism and Phenotypic Manipulation in Primates, Including Hypothesized Examples. Naturalistic or Captive Experiments with Animals are Required to Verify the Likelihood that Intraspecific Social Parasitism Explains these Responses (see text for further discussion).

Social parasitism and/or phenotypic manipulation	Possible examples in primates
Manipulation of parents by offspring (e.g., for time or energy) beyond resources required for survival	Animals, including humans (Trivers, 1972; Alexander, 1974); *Papio cynocephalus* (Altmann *et al.*, 1978).
Social parasite kills or effectively sterilizes host's (victim's) reproductive success (*social parasitoid* strategy)	*H. sapiens* (Miethe and McCorkle, 1998; Daly and Wilson, 1988)
Social parasites (often subordinates) exert toll on host's (often dominants) reproduction	*Pan paniscus* (Vervaecke *et al.*, 2003); some cases of infanticide (e.g., Crockett, 2003; Palombit, 2003); *Alouatta palliata* (Jones, 1997a)
Social parasites "rebel" against hosts	"Rank reversal" (e.g., Zucker and Clarke, 1998; Beekman *et al.*, 2003)
Ability of some individuals to manipulate others' mental state and/or behavior	*H. sapiens* (Frith and Frith, 1999); *Pan troglodytes* (de Waal, 1987; 1989)
Exploitation by social parasites of host's sensory biases [e.g., camouflage, mimicry or deception in various modalities (e.g., use of artificial attractants such as perfume)]	*H. sapiens* (Huggins and Preti, 1981; Preti *et al.*, 2003; Rodriguez *et al.*, 2000; Stern and McClintock, 1998; Dulac and Axel, 1995; Milinski and Wedekind, 2001; Wedekind *et al.*, 1995); *Papio hamadryas* (deceptive sexual swellings by females; Zinner and Deschner, 2000); *Alouatta palliata* (e.g., paedomorphic vocalizations by adult males; Jones, 1980, 1985a, 1997a); North American passerines (external appearance; Stokke, *et al.*, 2002); cichlid fish ("infantilization"; Shaw and Innes, 1980)
DNA mixing between host and parasite (e.g., by mating)	*A. palliata* "coercive mating"; Jones, 2002b; Jones and Agoramoorthy, 2003
Variation in virulence (e.g., aggression) of social parasite	*P. troglodytes* (de Waal, 1989); hominoids (Wrangham and Peterson, 1996); also see Thierry, 2000
Parental (maternal, paternal, or alloparental) manipulation of offspring	Primates (Hrdy, 1976, 1999a,b; Nicolson, 1987); *Macaca sylvanus* (Kuester and Paul, 2000); *H. sapiens* (Trivers, 1972; Alexander, 1974; Jones, 1986)
Dispersal rates of social parasites compared to hosts	*A. palliata* (see Glander, 1992)
Invasion of host kin groups by social parasites (e.g., by marriage, by dispersal)	*H. sapiens* (Alexander *et al.*, 1979); *Alouatta* (bisexual dispersal; see Brockett *et al.*, 2000a and references)
Genetic control of hosts by social parasites	*H. sapiens* (Manning *et al.*, 2002; Hager and Johnstone, 2003; Little *et al.*, 2003)
Interaction between social parasites (e.g., cooperation between subordinates)	Primates (see Smuts, 1987a,b)
Eavesdropping or otherwise behaving to obtain information about conspecifics	*H. sapiens* and other primates (see Dunbar, 1997; Whitfield, 2002)

It should be possible to disinhibit or extinguish responses resulting from ISP, then, unless ISP has imposed permanent changes upon the victim or host and thus, has had nonreversible effects. Theoretical and empirical tests of ISP may also be possible because analogous to studies of disease life-history (Day, 2003), social parasite virulence may be constrained by a tradeoff between mechanisms of influence by social parasites on hosts on the one hand, and host mortality induced by social parasites (i.e., social parasite virulence) on the other. Thus, following Day (2003), the logic, some of the terminology, as well as the methodology of the literature on aging should be applicable to theoretical and empirical investigations of social parasitism and phenotypic plasticity. Future research is required to determine the extent of intraindividual variation in these behaviors and their relationship to features of the abiotic and biotic (including social) environment (e.g., local competition).

Results from studies of social parasitism in insects and birds suggest that social parasitism and phenotypic manipulation should be considered as possible explanations for damaging behavioral events, particularly where the interactants are related (Parker and Rissing, 2002). Social parasitism may also influence the expression of social behavior (e.g., alliances or coalitions) in recipients of the parasitic or exploitative actions, as O'Donnell (1997) has shown for wasps (*Mischocyttarus mastigophorus*). Social parasitism and phenotypic manipulation, then, may explain behavioral phenomena previously explained by alternative hypotheses. For example, a highly influential treatment of animal signaling (Zahavi, 2003) assumes that signals are reliable ("honest") because individuals always behave in their own selfish interests by basing decisions upon benefits to direct reproduction. However, because of temporal and spatial environmental heterogeneity and its consequences, it may not always be in an actor's interest or it may not be possible to make choices on the basis of what is optimal for direct descendants. In some regimes, alternative tradeoffs (Day, 2003) may be favored such as preferring distant relatives, members of phenogroups, exploitation, or deception (see, for example, Otte, 1975; Byrne and Whiten, 1988; Whiten and Byrne, 1997; Jones, 1997a, 2002a; West *et al.*, 2002).

Negative Reinforcement as a Mechanism Of Exploitation in Heterogeneous Regimes

Jones (2002a) discussed the evolution of "negative reinforcement" as a set of postpunishment responses whereby a victim, opponent, or "client" might escape or mitigate the potentially deleterious consequences of interaction with a victimizer, cheater, or host. As Johnstone and Bshary (2002) show with game theoretical analysis, an interaction may switch from mutualism to parasitism as a result of the differential ability of a victim to avoid, escape, or terminate a conspecific's manipulation and/or exploitation (i.e., differential influence, power, and/or control). Importantly, Johnstone and Bshary (2002) show that a constraint on the evolution of parasitism will be costs in time imposed, potentially, by the host's attempts to avoid, escape, or terminate manipulation and/or exploitation. Thus, the victimizer's control is expected to be "incomplete" where costs in time outweigh potential benefits from social parasitism.

Negative reinforcement and other potential mechanisms of response to victimization (Mazur, 2002; Anderson, 1995) are expected to vary over time and space as optima change. These conditions are expected to favor the evolution of behavioral flexibility and a capacity for switching from one alternative to another depending upon condition-dependent tradeoffs (see, West-Eberhard, 1979; Jones and Agoramoorthy, 2003). Research is needed to document the conditions under which victims are capable of escaping potential or actual victimizers; however, it is expected that these environmental regimes will be most likely to be associated with limiting resources that are temporally and/or spatially ephemeral (Jones, 2000), conditions expected to disfavor resource monopolization by dominants, *ceteris paribus* (see Chapter 6).

"Individuality" as a Primate Signature Constraining the Evolution and Expression of Behavioral Flexibility and "True" Sociality

Paradoxically, individualized traits and patterns of trait combination can occur simultaneously with a broad range of facultative responses in an organism's behavioral repertoire. Crespi and Choe (1997a,b) point out that social parasitism may select for either increased or decreased levels of sociality, possibly dependent upon differential costs and benefits of social defense (and/or foraging?). Social parasitism, then, should be investigated as a possible selection pressure for the evolution of individuality in certain regimes, a force expected to retard cooperative and altruistic inclinations (Jones and Agoramoorthy, 2003, pp. 128–130) since individuality is expected to favor selfish (and, possibly, spiteful) responses.

Individuality (e.g., personality; Gosling *et al.*, 2003; see Maestripieri, 2003a,b) is a hallmark characteristic of many primates and represents a capacity to employ the same response patterns over and over again in temporally or spatially heterogeneous regimes. Similar to the effects of specialization, this type of response effectively increases the within-individual component of a population's total niche width, which Bolnick *et al.* (2003, p. 3) define as "the variance of total resource use of all individuals" and which these authors point out will be a function of sex, age, morphology, and diet. Like specialization, individuality decreases niche overlap between individuals, thereby decreasing interindividual competition and leading to preference for a particular subset of (limiting) resources (Bolnick *et al.*, 2003). Similar to investigations of specialization (Bolnick *et al.*, 2003), individuality may be considered in relation to a population's "realized niche"—variations in resource use as a function of intrinsic (e.g., morphology, behavior) as well as extrinsic (e.g., resource patchiness, interaction rates) factors. To the extent that individuality generates "consistent patterns in the way individuals behave, feel, and think" (Gosling *et al.*, 2003, p. 256; Maestripieri, 2003a,b), it is expected to retard and/or constrain the expression of behavioral flexibility and, further, the evolution of "true sociality" among primates (Jones and Agoramoorthy, 2003, pp. 128–130; Kitchen and Packer, 1999).

Although individuality may actually decrease the range of responses expressed by an individual in many conditions, the topic is essential to an understanding of behavioral flexibility in primates by highlighting that individualized responses contribute to the overall diversity of an individual's behavioral repertoire, that the frequency, rate, duration, intensity, and type of individualized response patterns signal information important to the optimization of benefits to genotypic and/or phenotypic success, and that individualized ("consistent") responses involve some sort of switch mechanism from less individualized to more individualized responses. Further, since patterns of signaling may mark individual identity and facilitate discrimination among conspecifics (Parr and de Waal, 1999; Bergman *et al.*, 2003), the resultant "consistency" in the communication of identifying cues or signals may, in part, be responsible for the evolution of complex social behavior in the Primate Order (Mendl *et al.*, 2002) and may represent a signature of social evolution in primates and some other social mammals (e.g., dolphins).

Behavioral flexibility, then, combined with "consistent" cues and signals may imply the employment of a broad range of responses in temporally and spatially heterogeneous conditions as well as the combination and recombination of a set of signals for purposes of communication with conspecifics constrained by individualized markers ("personality"), perhaps yielding a large, condition-dependent repertiore or vocabulary through a variety of sensory modalities (Hurd, 1997; Nur and Hasson, 1984). While such proliferation of signals would render individuals subject to manipulation, exploitation, and, possibly, control by social parasites, as suggested above, it is expected that individuality will be favored where the benefits to fitness outweigh the costs, on average. It is important to note that, in the insect literature, social parasitism has generally been studied as the parasite's successful decoding of the host's communication system (Hölldobler and Wilson, 1990), a relationship in need of investigation in primates and other social vertebrates.

In some environmental conditions it is expected to benefit individuals to identify phenotypes similar to or different from themselves (Sinervo and Clobert, 2003; Baglione *et al.*, 2003; Dickinson and Koenig, 2003). Such decisions may be made by "self-referent phenotype matching," may represent "green beard" effects, or may occur as a result of other mechanisms.

Individuals may positively or negatively assort as *phenogroups*, where it is, on average, advantageous for them to do so, and individuality may facilitate or inhibit genotypic and/or phenotypic success, depending upon the differential advantages and disadvantages to an individual of positive phenogroup assortment on the one hand, and negative phenogroup assortment on the other (see Jones, 2003a). Social parasitism, social learning, or other mechanisms may employ individual traits ("individuality") as markers for copying (e.g., conformity), resulting in the creation of new phenogroups, a potentially dynamic and creative source of innovation within societies, a process in need of investigation for primates and other social vertebrates.

Conclusions

The use of individual characteristics as markers for decision-making influencing the expression of alternative behaviors in varying conditions may be a function of the differential costs and benefits to individuals of positive assorting, interindividual (e.g., cooperative) vs. solitary tactics and strategies for the optimization of fitness. While it is tempting to suggest that females are more likely to be influenced by individual effects than males because they are expected to be more dependent than males upon relationships (especially relationships with kin; Wrangham, 1980; Sterck *et al.*, 1997), it seems premature to draw this conclusion until more is known of the discriminative capacities, tactics, and strategies of primate males and females (Chapter 4). Subsequent chapters discuss female (Chapter 5) and male (Chapter 6) behavioral flexibility in relation to age, dominance rank, and environmental heterogeneity as it may pertain to constraints imposed upon the sexes in an attempt to advance an understanding of the range of responses available to males and females under a variety of regimes.

Social Cognition and Behavioral Flexibility: Categorical Decision-Making as a Primate Signature

4

> For each individual primate, [group living] sets up an environment favouring the use of *social manipulation* to achieve individual benefits at the expense of other group members....
>
> Byrne and Whiten, 1997, p. 2 (emphasis in original)

Introduction

Social cognition may be incorporated within the Triversian (Trivers, 1985) meaning of sociality whereby interindividual interactions are classified as selfish, cooperative, altruistic, or spiteful. Like all social behavior (Frank, 1998; Trivers, 1985), I assume that social cognition has evolved to optimize fitness where the (genetic and/or phenotypic) interests of individuals competing for limiting resources are not equivalent. These interests might be compromised, however, where cognitive responses have been manipulated or exploited so that individuals behave in the interests of others (e.g., as victims of social parasitism; see Frith and Frith, 1999; Whiten and Byrne, 1988; Chapter 3).

Social Cognition as a Generator of Behavioral Flexibility

Maynard Smith and Price (1973) showed that damaging responses are expensive in *energy* (high rates of energy investment per unit time), while

47

nondamaging responses are expensive in *time* (low rates of energy investment per unit time). Like other forms of social response (Fagen, 1980; Parker, 1974), the consequences of social cognition can be divided broadly into these two categories: damaging (i.e., escalating or life threatening, such as homicide) and nondamaging (nonescalating, such as negotiation), outcomes which can be combined and recombined to yield flexible responses. Where advantageous, tactics and strategies of social cognition may also be employed to increase unpredictability relative to the perceived characteristics of an opponent's behavioral patterns and intentions (Miller, 1997). These conditions may arise wherever it is an actor's interest to confuse or to deceive a conspecific (see Jones, 1997a). Social cognition, then, may be employed to manipulate and/or exploit the differential costs and benefits of interactions with conspecifics.

Although tactics and strategies of social cognition *by the potential signaler* may lead to either damaging or nondamaging behavior *by the recipient of signals,* neural mechanisms of social cognition involving higher-order reasoning or planning may represent a repertoire of processes more costly in time than energy—compared, for example, to brain responses based upon emotion or impulse. Following the predictions in Figure 1.1, tactics and strategies of social cognition requiring *categorization* (Fig. 4.1) are expected to be associated with responses exhibiting relatively low likelihoods of escalation to damaging behavior (e.g., fighting; Parker, 1974). Categorization may be a primate signature representing a significant evolutionary transition because the process is thought to be related to self-perception as well as perception of others, possibly permitting an organism to perceive causality and to possess a "theory of mind" (see Barsalou, 1992; Maestripieri, 2003; Premack, 2004).

Similar to the use of conventional and ritualized displays, tactics and strategies of social cognition may be employed to optimize the benefits or to minimize the costs of interactions with conspecifics where potential signalers assess their future gains from damaging behavior to be lower than their costs. Mechanisms of social cognition (e.g., discrimination and categorization; Crowley, 2003) may also be employed where the future benefits from nondamaging behaviors outweigh the benefits of damaging interactions (e.g., fighting; Fig. 1.1). Employing a standard choice task common in psychological studies, Brosnan and de Waal (2003), for example, recently provided evidence suggesting that capuchins (*Cebus apella*) are capable of a simple form of mind-reading whereby "fairness" is judged by assessing costs and benefits from each interactant's point of view. While these preliminary results require replication, they indicate that nonhuman primates may be preadapted for the types of tactics and strategies of social cognition characteristic of humans, a claim made by these researchers. Tactics and strategies of social cognition (e.g., mechanisms for understanding the intentions and agency of conspecifics) may be employed in response to the range of potential interactions with competitors, opponents, victims, or "clients" (e.g., Fig. 1.1) and, in competitive contexts

Fig. 4.1. For these fraternal twins and their older sibling, kinship may or may not facilitate cooperation, probably dependent upon local conditions, especially competition for limiting resources, often parental ones. Similarities and differences in traits are expected to be employed in decisions pertaining to negative or positive phenogroup assortment, a process of assessment that may be facilitated by categorization. The potential for cooperation among kin or the expression of other social behaviors is expected to vary over time and space, and Johnstone and Roulin (2003) show that it may benefit siblings to employ negotiation in order to avoid potentially damaging interactions. ©Joshua L. Palmer.

and are likely to be utilized to minimize the likelihood of costly damaging fights.

Silk *et al.* (2003), for example, studying baboons, recently showed that "sociality enhances the fitness of nonhuman primate females" (p. 1233). Defining sociality as the amount of time a female is groomed by other adults in her group, these authors argue that grooming may be utilized "when social conditions are unstable" (p. 1233), implying that this behavior might be utilized where uncertainty prevails—like those conditions thought to favor categorization and mixed (probabilistic) tactics and strategies such as states requiring assessment in which costs and/or the likelihood of error are high (Crowley, 2003; Plaistow *et al.*, 2004). If grooming is one strategy to manage interindividual conflicts of interest, then it can be analyzed within the framework of "competition dependent" responses favoring behavioral flexibility and mixed strategies where interindividual interactions create heterogeneous conditions (Krebs and Dawkins, 1984; Miller, 1997). Alternatively, grooming may be employed as appeasement functioning as social parasitism (see Lenoir *et al.*, 2001) whereby the groomer manipulates and/or exploits the phenotype of the groomee (see Brockett *et al.*, 2000b, Fig. 2).

Mechanisms to understand the intentions and agency of conspecifics, then, may be interpreted within the framework of interindividual competition for limiting resources (e.g., food, mates, space, influence) as tactics and strategies to minimize the costs of competition. As such, tactics and strategies of social cognition (e.g., categorization) may be employed not only to defeat but also to manipulate or exploit conspecifics, and opponents may be victims of each other's cognitive capacities. It is in this sense that complex sociality may select for higher-order cognitive mechanisms to optimize fitness in those taxa preadapted to respond in a conventionally intelligent manner to environmental challenges.

The Costs and Benefits of Damaging and Nondamaging Outcomes of Social Cognition

Most discussions of interindividual conflict in primates have been limited to agonistic action patterns, and few of these reports discuss mechanisms of social cognition within a Hamiltonian framework. In a general theoretical paper, Maynard Smith and Price (1973) considered only escalating and nonescalating (e.g., ritualized) agonism, though Fagen (1980), by inference, extended their conclusions to all forms of damaging or nondamaging social behavior. The present chapter considers mechanisms of social cognition (e.g., categorization) within the context of these papers and assumes that tactics and strategies to understand the intentions and agency of conspecifics can be analyzed successfully within the parameters of the Hamiltonian unification program. Maynard Smith and Price (1973), for example, demonstrated formally that the costs of damaging and nondamaging conflict differ—the former entailing costs to fitness associated with fighting ability, the latter entailing costs in time (e.g., persistence, Alberts *et al.*, 2003 or persuasion, Jones, 1997a; see Chapter 3), and it seems reasonable to assume that tactics and strategies of social cognition represent mechanisms evolved to minimize the costs of aggression (after Maynard Smith, 1974; Parker, 1974; Krebs and Davies, 1993).

Persistence may be particularly characteristic of the tactics and strategies of social cognition as exemplified by the queuing behavior of male baboons (Alberts *et al.*, 2003), a strategy decreasing the likelihood of damaging behavior between subordinate and dominant males. Queuing is expected to be a time-dependent strategy requiring categorization to minimize costs where individuals interact repeatedly and the potential for damaging aggression is high (see Crowley, 2003). Like conventional and ritualized behaviors and similar to nondamaging responses, social cognition is here proposed as a set of higher-order competitive responses employed where low rates of energy expenditure per unit time are favored. Indeed, Zayan and Vauclair (1998, quoted in Crowley, 2003) argued that categorization, one mechanism of social cognition, may reduce "cognitive demand." Since the conclusions of Maynard Smith and Price (1973) may be generalized to all social responses, I conclude that social cognition entails significant costs in time to reduce energetic or other costs of decision-making mechanisms. Interestingly, Heinze and Keller (2000) have suggested that eusociality has evolved as an energy-saving strategy, an idea that might be generalized to other social transitions such as the transition from discrimination to categorization analyzed by Crowley (2003).

Costs associated with the expression of responses entailing a high rate of energy investment per unit time (e.g., escalating behavior) are expected to increase geometrically with increased time investment, presumably because these responses involve rapidly increasing rates of energy expenditure (i.e., increasing rates of metabolism, caloric output, and heat generation per unit time). The expression of responses entailing a relatively low rate of energy investment per unit time (e.g., mechanisms of social cognition), however, involve costs increasing linearly, all other things being equal, reflecting a relatively steady rate of increase in cost as time investment increases. Thus, consistent with Maynard Smith and Price (1973), the costs of responses entailing high and low rates of energy expenditure per unit time differ, and an increased investment in the former may be viewed as increasing response intensity (e.g., escalation to fights), in the latter, as increasing response frequency (e.g., increased numbers of interactions such as persistence or persuasion; see Brown *et al.*, 1997; McCleery, 1978; West-Eberhard, 2003, Chapter 23).

Where asymmetrical contests obtain, individuals with the greatest resource holding potential (RHP), usually dominants, may employ nondamaging responses [i.e., responses with low rates of energy expenditure per unit time (e.g., displays or mechanisms of social cognition) to advertise likelihoods or thresholds of escalation to individuals of lower fighting ability (usually subordinates)]. These individuals win most competitive interactions based upon asymmetries (e.g., in strength, size, age, sex; Parker, 1974; Dawkins and Krebs, 1978). Responses requiring high rates of energy expenditure per unit time are rarely observed in nature (e.g., "principle of stringency"; Wilson, 1975). The hypothetical benefits to an individual with greater RHP than his/her opponent are greater, on whole, than those to his/her opponent. This condition is likely to obtain in part because, all other things being equal, initial costs to the individual with greater RHP rise more slowly, presumably because of a more efficient or prudent utilization of time and energy. Apparently, it will benefit the former individual to advertise his/her readiness to fight for disputed resources where his/her benefits from winning a conflict are expected to be greater than those of his/her opponent (Maynard Smith, 1974; Parker, 1974).

It is important to note that, in the same conditions, the costs of escalation are higher for the individual utilizing tactics and strategies employing low rates of energy expenditure per unit time than for the individual utilizing tactics and strategies employing high rates of energy expenditure per unit time, all other things being equal. Selection pressures for energy saving tactics and strategies (e.g., tactics and strategies of social cognition) are expected to be most intense upon subordinates (West, 1967; West-Eberhard, 1979; Jones, 1996a) where these individuals have more to gain than to lose by employing these mechanisms. If the individual with lower RHP or the individual likely to experience greater uncertainty or risk avoids, escapes, or terminates a potentially damaging encounter after his/her opponent expresses overt aggression but before (s)he has responded likewise, his/her costs will be lower than those of the individual prepared to utilize high rates of energy investment per unit time (e.g., damaging aggression) to resolve the conflict. The potential escalator, then, should escalate with prudence (Jones, 2002a).

Further, as Maynard Smith (1974) makes clear, the costs of nonescalating encounters are not only lower than those for escalating encounters, but are also relatively equivalent for both opponents (competitors). In fact, it may cost less to reciprocate social behavior with low likelihoods of escalation (see Fig. 1.1) than to withdraw, escape, or avoid interactions, particularly if these tactics and strategies increase the likelihood of interference or other costs by a competitor in the future. These conditions will have significant evolutionary effects and will benefit individuals utilizing mechanisms employing low rates of energy investment per unit time (e.g., mechanisms of social cognition such as categorization).

Competitive Behavior and Resource Dispersion Related to Social Cognition

Discussions of damaging and nondamaging behavior have usually focused on patterns of relative status or rank that may result where individuals engage in contest competition for a clumped, limiting resource (Emlen and Oring, 1977). In these conditions, escalators are presumed to receive "priority of access" to resources (after Hausfater, 1975; Popp and DeVore, 1979). Damaging behavior is likely to be associated with particular patterns of resource distribution and abundance in time and space, and individuals with greater RHP will have the greatest advantage where resources are limited, clumped, and ephemeral and where competition is direct rather than indirect. However, where the distribution of a resource can be manipulated (e.g., where males compete for mates and "herd" females; Wittenberger, 1980; Jones, 1985b), the competitive disadvantages to nonescalating males might be minimized (Trivers, 1971), perhaps with tactics and strategies of social cognition (e.g., "mind-reading," deception; Kummer, 1968; Essock-Vitale and Seyfarth, 1987; de Waal, 1989).

Interindividual Conflicts of Interest as a Selective Pressure for Mechanisms of Social Cognition

Discussions of the functions of dominance hierarchies imply that they reflect cooperation among group members, that subordinate individuals "settle" for low status rather than risk damaging encounters, and that dominant individuals "altruistically" tolerate the presence of subordinates who consequently obtain a share of "goods and services" (Fagen, 1980). It follows from the previous discussion, however, that it may be especially beneficial to individuals with lower RHP than their opponent (i.e., individuals most likely to utilize tactics and strategies with low rates of energy investment per unit time such as persistence or other tactics and strategies of social cognition) to employ nondamaging responses (e.g., mechanisms of social cognition) in some competitive regimes. Dawkins and Krebs (1978) point out that "as an inevitable byproduct of the fact that animals are selected to respond to their environment in ways that are on average beneficial to themselves, other animals can be selected to subvert this responsiveness for their own benefit" (p. 285). Individuals, then, may be expected to behave so as to influence the escalation likelihoods of their competitors with energetically prudent responses (Jones, 2002a; Krebs and Davies, 1993; Popp and DeVore, 1979). Grooming is a good example of a response that is unlikely to escalate and that may be employed by subordinates or dominants. A subordinate grooming a dominant appears to communicate, "I'll scratch your back; please scratch mine." Where dominants have been observed to groom subordinates (e.g., *Alouatta palliata,* Jones, 1979; *Cebus apella,* Parr *et al.,* 1997), the message might be, "I'm scratching your back; you'd better scratch mine."

Parker (1974) discussed "asymmetries" other than fighting ability, which may obtain between competitors. Where a resource is of greater benefit to one competitor than to another, it will "pay" the former to expend more energy in winning the encounter (see, for example, Jones, 1980, p. 400). Thus, if $E_g < N_g$, where E_g equals an escalator's (e.g., a fighter's) gain in fitness from winning a competitive interaction and N_g equals a nonescalator's gain in fitness from winning, the nonescalator will have the advantage. A nonescalator may invest her/his total "fitness budget" (Parker, 1974) in nondamaging behavior (e.g., grooming), adopting a "pure" or a "mixed" strategy of conventional (e.g., displays) and/or nonagonistic responses (Dawkins and Krebs, 1978). Such investment may effectively increase the aggressive costs of an escalator by inflicting additional costs in time, thereby minimizing differences in status or rank between individuals and potentially altering balance of power or influence within dyads. This condition obtains where costs incurred by an escalator serve to effectively *decrease* the value of a resource to him/her.

The nonescalator's potential benefits from social behavior may also be optimized by a paradoxical effect suggested by Popp and DeVore (1979). If an escalator continues to respond to a nonescalator's signals with escalating behavior, the former communicates the "intent" to escalate to damaging aggression. The escalator will *prefer* to maintain his investment at a threshold where his benefit to cost tradeoff is maximized. A nonescalator, however, should be particularly sensitive to the asymmetry between dominant and subordinate, a state which may favor the nonescalator in some conditions (e.g., Strier *et al.*, 2002; Chapter 6).

Thus, a nonescalator may influence an escalator's defensive costs and benefits, perhaps with tactics and strategies dependent upon mechanisms of social cognition (e.g., deception, manipulation, superior knowledge, "emotional intelligence," or other intelligent, exploitative, or clever responses). A nonescalator will be required to decide when it will pay him/her to expend effort to induce an escalator to increase his or her investment in social behavior. The ideal tactic or strategy for the nonescalator would be to employ these responses in such a manner that the escalator would expend energy unnecessarily. A behavioral strategy of "bluff," deceit, and/or "testing" (Zahavi, 1974; Brown, 1978) or, perhaps, a compound display (see Chapter 6) might create a condition of "social neglect" for the escalator whereby (s)he is induced to invest in social responses to his/her disadvantage but to the advantage of his/her competitor. This strategy avoids the pitfalls of spite if potential benefits to the individual with lower RHP are greater than potential benefits to his/her opponent (Vickery *et al.*, 2003; Alberts *et al.*, 2003). Consistent with this view, Chapter 6 proposes that future studies of adult males competing for mates in the same social unit should characterize the differential tactics and strategies employed by males as a function of their rank distance.

Any heterogeneous biotic or abiotic factors (e.g., variations in habitat quality, predation, demography; McNab, 1980) that may increase the costs or decrease the benefits of damaging behavior to an escalator will work to a nonescalator's advantage. Thus, ecological and life historical conditions permitting, it will always "pay" a nonescalator to increase his investment in nondamaging social interactions with an escalating competitor in these regimes. This condition will hold even where an escalator "decides" to employ damaging behavior if the nonescalator can avoid, escape, or terminate a fight (Jones, 2002a).

Formal models and empirical data are required to assess the utility of the continuum approach implied in Figure 1.1 and inferences from the previous discussion. However, some available reports (Kummer, 1968, 1995; Hausfater, 1975; Jones, 1980) suggest that nondamaging responses have been favored for the resolution of interindividual conflicts of interest and the purchase of "goods and services" where the costs of escalated aggression are very high or difficult to estimate. The use of nondamaging tactics and strategies as competitive responses might be particularly beneficial to subordinate individuals, to individuals in heterogeneous environments, and to individuals of species with long lifespans where fitness optima are expected to change over time and space. An understanding of nondamaging responses may also facilitate the interpretation of contexts in which "priority of access" fails to correlate consistently with measures of dominance rank.

Resolving Conflicts of Interest with Probabilistic Responses

Although some conditions may favor individuals fighting to the point of death to resolve interindividual conflicts of interest (West-Eberhard, 1979), tactics and strategies to avoid damaging aggression have arisen in most species (Krebs and Davies, 1993; Chapter 6). One outcome of these responses may be dominance hierarchies, common in primate taxa, particularly among males (Walters and Seyfarth, 1987; Dixson, 1998). A frequent explanation for the expression of social structure, including dominance hierarchies, is the mutual assessment of RHP by individuals who interact repeatedly and are able to remember past events. Most attempts to provide theoretical descriptions of these encounters have involved the use of evolutionarily stable strategy (ESS) models measuring variations in fitness between interactants of the same species adopting similar or different behavioral responses. ESS models describe individuals making decisions based upon proximate conditions (Jones, 2004; Jones and Agoramoorthy, 2003; see Chapters 1 and 2).

Because interactions are assumed by most ESS modelers to occur between individuals demonstrating *asymmetries* (inequalities) in RHP (see, for example, Leimar and Enquist, 1984), it has been generally claimed that outcomes will be based upon *discrimination* of differences leading to rapid assessment and "pure" or deterministic tactics and strategies. *Categorization* requires that certain distinctions between individuals be ignored (Crowley, 2003), and these responses reflect "mixed" or probabilistic tactics or strategies. It is assumed by most authors (Gross, 1996; Austad, 1984) that mixed strategies are unlikely to arise because of presumed asymmetries between opponents, but recent theoretical formulations show that mixed strategies resulting from probabilistic behavior can be evolutionarily stable where it benefits the individual to recognize "equivalence classes" by ignoring differences (Crowley, 2003), for example, where the discrimination of asymmetry is uncertain or risky. In these conditions, categorizing rather than discrimination is likely to be expressed. Categorization will obtain *within* asymmetrical classes (e.g., age, sex, size) and/or where RHP asymmetries are small, ineffectual, dangerous, or difficult to assess, for example, where "psychological" factors vary between individuals (e.g., motivation, emotion, intelligence). In these conditions, discrimination based upon assessment of differential RHP may be costly.

Crowley (2003) shows, in addition, that categorization will be favored not only where asymmetries between serially interacting conspecifics are small and/or uncertain but also where asymmetries are moderate *if individuals interact frequently*, as would be expected for group-living organisms (e.g., most primates). This author's formulations demonstrate the importance of studying responses between individuals as a function of relative rank differences, in particular, individuals whose rank distance is small since this condition may be especially prone to facilitating categorical decision-making.

Categorization as a Primate Signature

Primate phenotypes are characterized by behavioral flexibility (Miller, 1997; Smuts *et al.*, 1987; Runciman *et al.*, 1996; Boesch *et al.*, 2002; Jones, 2003b; Maestripieri, 2003a; Premack, 2004), suggesting that mixed strategies may be common in the order and that *categorization* may be a primate signature. Categorization implies intelligent behavior although not all intelligent behavior implies categorization. The impressive capacity of Japanese macaques to categorize dominance rank on the basis of matrilineal membership (Chapais *et al.*, 1995) is intelligent if the response is rule-governed and not an elementary result of conditioning. While most cases of discrimination learning can be explained with a straightforward S → R (stimulus, response) model based on operant conditioning (Mazur, 2002, pp. 227–242), categorization may require a S → O → R (stimulus, organism, response) model implying concept formation (Mazur, 2002, pp. 242–249).

Animal (including human) studies of concept formation were investigated by Rosch (1975) who described the architecture of "natural" categories—categories of events encountered in the real world (e.g., plants, animals). Rosch's studies with humans and other research with nonhumans (e.g., Herrnstein and Loveland, 1964) demonstrated that the classification of natural events is complex, requiring more than purely associative abilities. This body of research documented the relative ease with which humans and many other organisms learn natural categories, and additional studies showed that humans (Sidman and Tailby, 1982) and, possibly, other animals (chimpanzees, Yamamoto and Asano, 1995, Fig. 4.2; pigeons, Vaughan, 1988) are capable of *stimulus equivalence* whereby individuals learn the interchangeability of stimuli with imperfect information (Mazur, 2002, p. 248). Stimulus equivalence may be a precursor to categorization where certain information must be ignored in order to form functional classes and may be a necessary condition for the evolution of flexible behaviors dependent upon tactics and strategies of social cognition.

Studying baboons (Fig. 3.1, Fig. 4.3) without language learning, Bergman *et al.* (2003) have recently demonstrated the ability of these animals to classify

Fig. 4.2. Chimpanzees (*Pan troglodytes*) like this adult male may be capable of stimulus equivalence after language training (Yamamoto and Asano, 1995). The ability to construct equivalence sets may be involved in categorical decision-making where some information is ignored in the formation of functional classes. ©Mike Seres.

Fig. 4.3. Anthropoids, such as these baboons (*Papio cynocephalus*), may demonstrate higher-order capacities of social cognition which may permit categorization and consequent mixed strategies. These mechanisms may represent ancient primate traits. ©Mike Seres.

conspecifics according to their "membership in higher order groups" (p. 1234) and other individual attributes. These studies suggest that baboons can discriminate relations both between and within categories and emphasize the significance of novel stimuli as markers of categories or classes. While it is unclear to what extent capacities such as those demonstrated by Bergman and his colleagues represent traits shared by other social mammals (Dunbar, 1998, 2003; Connor *et al.*, 1999), it seems likely that higher-order categorization "may have quite ancient roots among the primates" (Dunbar, 2003, p. 1161; see Tomasello and Call, 1997; Dunbar, 1998).

Conclusions

Many primates are obligately social and interact for extended periods of time with other group members of the same or different age or sex. These conditions create contexts in which individuals might be favored who deceive, exploit, or manipulate the phenotypes of others (Jones, 1986, 1997a; Jones and Cortés-Ortiz, 1998; Bronstein, 2001, 2003; see Chapter 3), responses that may lead to counteradaptations for the prediction of conspecifics' behaviors, emotions, and/or thoughts which Miller (1997; also see Krebs and Dawkins, 1984) classifies as concealment of intentions, deception, and unpredictability. Psychologists have historically been concerned with concealed responses (e.g., emotions, thoughts), topics of study that led to the rise of behaviorism ("You cannot study what you cannot see.") within the discipline at a time when concealed responses could not be investigated empirically. Advances in technology (e.g., fMRI), however, have afforded psychologists the opportunity to study these phenomena with robust results (Aureli and Whiten, 2003). Discussing "the behavioral flexibility typical of human and nonhuman primates, especially during social interaction," Aureli and Whiten (2003, p. 318) propose that emotions may "mediate" (i.e., function as an intervening variable of) these responses.

Despite the classic texts by Byrne and Whiten (1988; Whiten and Byrne, 1997), little empirical work has been conducted on the topics of concealment, deception, and/or behavioral unpredictability among primates. Nonetheless, Miller (1997) argues persuasively that these traits, in addition to concealment of intentions, are diagnostic features of primates. The paper by Crowley (2003) discussed in the present chapter assesses one cognitive feature, categorization, that may be required for higher-order decision-making leading to concealment, deception, and unpredictability. What seems clear from available evidence (Maestripieri, 2003a) is that for humans, some other primates (e.g., *Pan, Papio, Cebus*) and, probably, other social mammals (e.g., Connor *et al.*, 1999, 2001; Kitchen and Packer, 1999, Fig. 9.4), certain environmental conditions preadapted these taxa for higher-order processing rather than similar operations based upon alternative mechanisms [e.g., obligate task specialization (Wilson, 1971) or "semiochemical" communication (Lenoir *et al.*, 2001)]. Categorization may be one such preadaptation originally favored by abiotic (e.g., climate) or biotic (e.g., food or social dispersion and quality) heterogeneity.

Female Primates as "Energy-Maximizers" in Heterogeneous Regimes

5

> [P]olygamy in social mammals is largely a consequence of selection favoring sociality among females.
>
> Wittenberger (1980, p. 197)

Introduction

The problem of any sexual organism is the optimization of genotypic and phenotypic benefits, and the problem of the female is to do her best reproductively within the energetic constraints of her local regime. All other things being equal, a female will be selected to invest the largest possible share of her total energy budget into reproductive activities (especially, for female mammals, parenting effort), rather than on growth and survival or mating effort. It is for these reasons that in many environmental conditions it will benefit females to exhibit highly selective tactics and strategies that are not energetically wasteful. As Alexander *et al.* (1997; also see Reeder, 2003) have pointed out, it is to the female's advantage to control events surrounding fertilization, and one of the possible benefits of multiple mating and/or cryptic female choice to primate females may be that physiological mechanisms would generally be less energetically costly than behavioral ones for purposes of control (see Holland and Rice, 1997). Shuster and Wade (2003, p. 131) point out, further, that "sperm competition and cryptic female choice can enhance or diminish the sex difference in strength of selection depending upon how the male variance in reproductive success is affected," an outcome that might benefit females in some regimes. Other mechanisms may also be favored in females. For example, multiple mating may enhance the immunocompetence of offspring (Johnsen *et al.*, 2000; Foerster *et al.*, 2003), and food quality may influence the timing of births (Miller, 2003; Brockett *et al.*, 2000c).

Discrimination Abilities, Allocation Strategies, and Behavioral Flexibility of Female Primates

Females will usually be capable of discriminating their mothers, their mother's offspring, their own offspring, and their siblings' offspring, although females are expected to be less capable of discriminating their full-sibs or, in the presence of multiple mating, the fathers of their offspring (see West-Eberhard, 1975). Females, all other things being equal, will experience relative confidence in their assessment of r_{xy} and r_{xe} (Chapter 2; see, for example, Bergman *et al.*, 2003), and the utility of behavioral flexibility for this sex is expected to be especially elaborate in relation to responses influencing parenting effort, including feeding. Box (2003) discusses evidence showing that female marmosets and tamarins exhibit a noteworthy degree of behavioral flexibility related to foraging and ecological challenges, although these results appeared to be correlated with age as well as gender since behavioral flexibility of subadult females was most pronounced.

In many other conditions (e.g., those involving mating effort), behavioral flexibility may represent a waste of energy for females, exposing them to increased likelihoods of escalation, to increased exposure to coercion, force, persuasion, or to social parasitism. It is for these reasons (i.e., the potential for wasted energy) that females may often benefit from adopting passive and/or indirect tactics and strategies (see Johnstone and Keller, 2000; Watson *et al.*, 1998; Smuts, 1987a,b; Smuts and Smuts, 1993; Soltis *et al.*, 1997a,b). These ideas are consistent with the view that, compared to males, females may be influenced more by stabilizing selection (see Lerner, 1970; Stearns, 2002) and that, all other things being equal, females are expected to be "energy maximizers" relative to males in the same conditions (Schoener, 1971).

It may not be the case that female strategies are passive but that it often pays females to surrender or to adopt *white flag* ("I give up!") tactics and strategies (see, for example, Johnstone and Keller, 2000; Watson *et al.*, 1998). Since both males and females "exist to reproduce" (Charnov, 2002) in the sense that every act has the potential to affect fitness, environmental heterogeneity may often lead females to limit their behavioral flexibility and/or to adopt energy-efficient responses compared to males in the same conditions. Such responses may in part explain the complex reproductive physiology of mammalian females compared to males as well as incipient (Jennions and MacDonald, 1994; Jones, 1996a; Solomon and French, 1997; Abbott *et al.*, 1998; Porter, 2001; Clutton-Brock *et al.*, 2003) or "true" (Wilson, 1971; Sherman *et al.*, 1991) division of labor (Sherman *et al.*, 1991) by females, all of which may be energy-saving strategies in some species in some conditions (Heinze and Keller, 2000; Jones and Agoramoorthy, 2003, pp. 124–125).

Such a view does not reject the idea that, in some environmental regimes, females may exhibit highly creative and, possibly, energetically expensive responses (Hrdy, 1999a) to foraging for food, to male mating behavior, to female competitors, and in relation to offspring, particularly where limiting resources vary in dispersion and/or quality above some threshold level. However, it can be expected that primate and other vertebrate females will allocate energy to mating effort, growth, and survival judiciously and only to the degree that the inclusive (genetic and phenotypic) benefits from doing so outweigh the costs (Shahnoor and Jones, 2003, p. 20; Queller, 1997) or to the degree that such energetically expensive expression represents exploitation and/or manipulation by conspecifics (e.g., relatives, dominants). Where resources required for females to reproduce are clumped but unpredictable in space and time, energetically expensive responses by females (e.g., female–female competition for food and/or mates) might be favored, as has apparently occurred for female mantled howlers in Guanacaste, Costa Rica, who fight and chase at a relatively high rate compared to males in the same tropical dry forest regimes (Jones, 2000; Preutz and Isbell, 2000; Goymann *et al.*, 2001; Boinski *et al.*, 2002).

As Brockmann (2001) pointed out, "allocation strategies" (e.g., investment in male vs. female offspring or investment in alternative behaviors) have three common features. First, time or other resources are invested in two or more different activities with a common function. A female *Alouatta palliata*, for example, may fight as well as enter into a coalition with another female in order to obtain access to resources (Jones, 1980, 2000; Barrett *et al.*, 2002; Hrdy, 1999a). Second, individuals utilize "decision rules" (not necessarily conscious) about which responses are optimal in particular situations ("strategizing"; Chapter 4). A female mantled howler, for example, may search for food individually rather than forage with her group (Jones, 1996a, 1998). Finally, allocation strategies share a feature common to all choices—time or other resources invested in one activity obviates investing those resources in another activity. Resources invested in sons, for example, represent resources not invested in daughters, and this tradeoff may vary from condition to condition, depending upon varying optima. Allocation strategies have been discussed in relation to life history, foraging, reproductive, and social decisions (Brockmann, 2001). Because of the primacy of energetics for females, however, allocation strategies pertaining to foraging may drive life history, reproductive, and social tactics and strategies, all other things being equal, in the absence of exploitation or manipulation.

Behavioral flexibility implies that individuals experience different optima over time and space, favoring more than one response. As Brockmann (2001, p. 11) points out, this condition will occur "when fitness curves (gain curves) cross," making one response optimal in one condition or state, another in a second, explaining the differential expression of behaviors by, for example, age or reproductive condition, population density, season, or habitat. A "switch" from one state to another may be determined by density-dependence, frequency-dependence, and/or competition-dependence (e.g., interaction rates), processes that are likely to stabilize frequencies of alternative behaviors within populations (Crook, 1972; Brockmann, 2001; West-Eberhard, 2003; Jones and Agoramoorthy, 2003).

Relative Reproductive Value as a Determinant of Behavioral Flexibility

Individuals are expected to assess the potential costs and benefits of their social behavior in accord with r_{xy} and r_{xe}. West-Eberhard (2003) describes the *assessment hypothesis* whereby relative quality, dominance rank, status, and/or resource holding potential function as an indicator of relative reproductive capacity (West, 1967; West-Eberhard, 1979; female *Presbytis* [*Semnopithecus*] *entellus*, Hrdy and Hrdy, 1976; female *A. palliata*, Jones, 1996a; Fig. 1.2), and longitudinal work on savannah baboons (*Papio cynocephalus*) suggests that populations of wild primates are age-structured (Altmann *et al.*, 1996; see Fig. 3.1). In these formulations, an individual's position in a group relative to others acts as a marker or signal to others based upon the individual's relative reproductive value (RRV; Hrdy and Hrdy, 1976). Relative reproductive value, then, should predict an individual's relative investment in alternative behaviors, *ceteris paribus*.

Jones (1996a) argued that temporal division of labor may reflect RRV since division of labor based upon age or size may reflect the reproductive condition of individuals in social groups. In 1967, West proposed the general hypothesis that hierarchical relations may be advantageous to both dominants and subordinates and that individuals of low rank may be inferior reproductives who benefit genetically from associations with and contributions to reproductively superior individuals, assuming that dominance rank reflects social (reproductive) quality. Since increasing age or size eventually entails decreasing reproductive value (v_x, Fig. 5.1), where the relative contribution to future generations of an individual of a given age is quantified, several authors have noted that the display of social behavior, such as foraging behavior benefiting all members of a group ("social foraging"), should increase with age as the benefits from individual (selfish) reproduction decline (West-Eberhard, 1975; Hrdy and Hrdy, 1976). As individual reproductive value decreases, benefits (genetic or phenotypic) from assisting the reproduction of conspecifics (the "West-Eberhardian/Alexandrian" definition of social behavior) may increase because costs (genetic or phenotypic) of social behavior decrease with decreased benefits from individual reproduction.

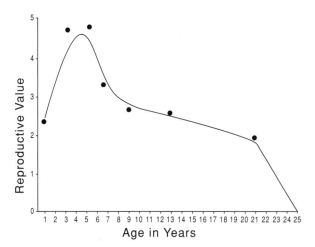

Fig. 5.1. The reproductive value (v_x) curve for the howler population at Hacienda LaPacifica, Cañas, Ganacaste, Costa Rica. Reproductive value was computed for the mid-point of each age class using equations cited in Jones, 1997b). The mid-point of each age interval was employed since discrete ages could not be determined year-by-year. Age in howlers is determined on the basis of tooth erosion where D1 (an individual approximately 5–7 years old) is the youngest adult class; D2 (about 7–10 years old) animals are middle-aged; D3 adults (aged 10–15 years) are middle-aged to old, and D4 adults (about 15+ years of age) are the oldest (see Glander, 1975). Tooth ages and chronological ages are presumed to correlate.

During an extended period of study at Hacienda La Pacífica, Cañas, Guanacaste, Costa Rica, Jones (1980, 2000) studied two marked, aged groups of mantled howler monkeys in two tropical dry forest habitats (riparian and deciduous). For this species, age and dominance rank are negatively correlated in the hierarchies of adults of both sex (Jones, 1978, 1980). In this study, foraging was operationally defined as the behavioral series: feed-rest-move (at least 100 m)-feed, by a unit of more than three adults. These criteria were adopted in order to standardize measurement and to eliminate periods of food search within unusually large patches and by consort pairs. Females who initiated foraging sequences were identified and results were analyzed by female age.

The null hypothesis held that the frequency of foraging by females of any age class would be proportional to the total number of females who foraged in an age class (Table 5.1). A monthly foraging rate for each forager was computed by dividing the frequency of foraging by the female's number of months resident in a group, a period of time varying from 10–14 months since some females emigrated during the study. These rates were compared with a female's age class on the one hand, and dominance rank on the other, to assess the relationship between the display of social foraging behavior and rank, and by inference, v_x.

Table 5.1. Age Class, Estimated Age in Years, Number of Females in Each Age Class (N), Observed (O) and Expected (E) Frequencies of Social Foraging, and Cumulative Chi Square (χ^2) Values for a Test of the Null Hypothesis (after Jones, 1996a).

Age Class	N	O	E	$(O-E)^2/E$
Young adult (5–7)	5	15	42.4	17.71
Middle-aged (7–10)	5	35	42.4	1.29
Middle-aged to old (10–15)	1	18	8.1	12.11
Old (15+)	1	33	8.1	76.54
Total	12	101	101.0	107.65

Table 5.1 presents Jones' (1996a) analysis for foraging frequency in the large riparian group (group 5) as a function of female age, including expected frequencies. Computing "goodness of fit" led to an unequivocal rejection of the null hypothesis ($p \leq 0.001$, $\chi^2 = 107.64$, df = 3). Thus, old age and foraging frequency are significantly related. Young adult females initiate foraging significantly less than expected on the basis of their numbers ($p \leq 0.001$), suggesting that such individuals are relatively "selfish" and/or are conserving time and, especially, energy, presumably for reproduction. Table 5.1 also shows that the middle-aged to old female foraged more than expected by chance ($p \leq 0.01$), and this female succeeded the oldest and lowest ranking female as the most frequent forager when the old female disappeared because of emigration or death in 1977 (C.B. Jones, personal observation).

Further supporting the results in Table 5.1, the relationship between foraging rate and age class (Fig. 5.2) yields a significant, positive correlation ($r_s = 0.629$, $p \leq 0.05$, N = 12). Related to this, the correlation between foraging rate and dominance rank (Fig. 5.3) is significant but negative (i.e., the higher the foraging rate, the lower the dominance rank, $r_s = -0.63$, $p \leq 0.05$, N = 12). Results for the deciduous forest group (group 12) support the findings for group 5 since the oldest female in this group socially foraged more frequently than any other ($p \leq 0.001$, $\chi^2 = 17.29$, df = 2). Again these findings show that the initiation of social foraging is significantly associated with female age and dominance rank and, by inference, with v_x suggesting that female *A. palliata* exhibit adaptive behavioral flexibility across their lifespan, foraging socially in response to differing optima at different ages. Indeed, extrapolating from Fig. 5.1, v_x and foraging rate are negatively and significantly correlated with foraging rate/mo ($r_s = -0.95$, $p \leq 0.02$, N = 12).

Jones' (1996a; also see Jones, 1996b) studies suggest that social foraging by old females is associated with ephemeral food (new leaves, flowers, and fruit), and an old female's presumed experience with the mosaic of her home range might enhance her efficiency as a forager so that her foraging

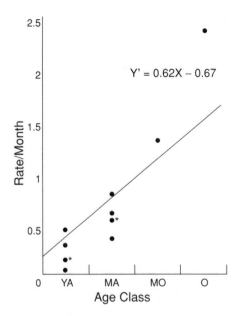

Fig. 5.2. Social foraging rate as a function of age for young adult (YA), middle-aged (MA), middle-aged to old (MO), and old (O) female howlers. Each point represents one adult female except where asterisk indicates two. Foraging rate was computed by dividing the frequency of foraging by the number of months the female resided in a group, a period of time varying from 10 to 14 months (Jones, 1996a).

activity may yield an energetic and nutritional gain to other group members, particularly relatives. Because social foraging may be viewed as a form of helping behavior, its costs and benefits are expected to vary as a function of local competition (Chapters 1 and 2). Temporal uncertainty of preferred food resources may favor individuals that are the beneficiaries of the foraging activity of others when v_x is low. Division of labor through differential social roles, common in primates, may be a function of RRV, and behavioral flexibility may be understood within the context of life history patterns. Both mantled howlers (Jones, 1980) and cooperatively breeding marmosets and tamarins (Abbott *et al.*, 1998) display numerous characteristics similar to social insects (e.g., temporal division of labor in *A. palliata*, Jones, 1996a; reproductive suppression in *Callithrix jacchus*, Abbott *et al.*, 1998; Schaffner and Caine, 2000), and both groups deserve intense investigation in order to document convergence in traits diagnostic of higher grades of sociality.

Behavioral flexibility as a result of RRV is likely to be more important for females than for males in the same regimes since assessment of coefficients of relatedness is likely to be more confident for females, variance in reproductive

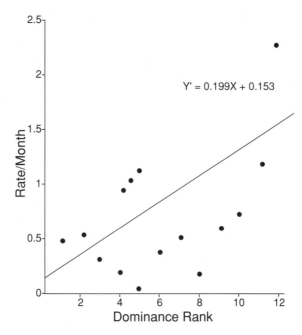

Fig. 5.3. Social foraging rate as a function of individual dominance rank. Each point represents one adult female. Note that low numbers represent high rank. Foraging rate was computed by dividing the frequency of foraging by the number of months the female resided in a group, a period of time varying from 10 to 14 months (Jones, 1996a).

success will generally be lower for females (with the possible exception of monogamous and cooperatively breeding species; Emlen and Oring, 1997; Andersson, 1994), and female fecundity will decrease with age to a greater extent than that of males in similar conditions. These observations may mean that behavioral flexibility in females may be more responsive to age-related changes compared to males who may be more responsive to ongoing and/or more exogenous conditions, especially competitive relations with other males for access to females. On the other hand, as the results reported previously indicate, females are expected to be responsive to ongoing changes in food dispersion and quality, exogenous heterogeneous effects that are expected to favor behavioral flexibility.

Behavioral Flexibility and Time-Scales Briefer than T

Behavioral flexibility in females may also be driven by time-scales briefer than presumed switch points sensitive to changes in age and reproductive condition, and female behavior may also be responsive to seasonal changes in climate and food dispersion and/or quality. *Alouatta* females, for example, may adjust the timing of reproduction in response to local humidity or the availability of fruit (Brockett *et al.*, 2000b). Brockett and her coworkers revealed patterns whereby female mantled howler monkeys (*A. palliata*) demonstrated similar patterns of birth for wet (riparian habitat of Costa Rican tropical dry forest and semideciduous lowland tropical forest at Barro Colorado Island, Panama, Clarke and Glander, 1984; Carpenter, 1934; Milton, 1982) and dry (deciduous habitat of tropical dry forest at two sites in Costa Rica, Jones, 1978, 1980; Fedigan *et al.*, 1998) sites. In wetter sites, researchers failed to find seasonal birth peaks, whereas in drier sites, dry season birth peaks were found, apparently corresponding to the dry season peak in fruit reported for deciduous habitat by Frankie *et al.* (1974). Consistent with these results, Crockett and Rudran (1987) reported a dry season peak in births for the Venezuelan red howler monkey (*A. seniculus*) in drier habitat. All of these authors provide support for the views that peaks in howler births are negatively related to rainfall and that the timing of births exploits food available to offspring at weaning. Results for the black and gold howler monkey in Argentina (*A. caraya*) support these conclusions (Kowalewski and Zunino, 2004; G.E. Zunino, personal communication; Brockett *et al.*, 2000b).

If drier conditions are likely to present with greater temporal and spatial heterogeneity than wetter conditions, all other things being equal, birth seasonality may be viewed as a flexible response by females to these regimes. Brown howler monkeys (*A. fusca clamitans*) in the Atlantic Forest of Brazil demonstrate no birth peaks despite pronounced seasonality of climate and resources in this habitat where peaks in fruiting occur during the rainy season (Strier *et al.*, 2001). Decisions to reproduce by female howlers, then, may be a result of some calculus of fruit availability, quality, and timing in addition to patterns of rainfall, and future research on howlers and other primates should investigate the permutations of these relationships that may be complex (multifactorial). Such studies may require methodologies rarely employed by students of primates (e.g., field experiments, including translocation experiments, Jones, 1982a; physical, including morphological, manipulations, Velando, 2002; "focal-tree" methods, Jones, 1983b). Manipulations can be justified ethically with populations or species that are not endangered, and it might be argued that field manipulations are endowed with a special urgency for the efficient collection of data to enhance our knowledge of disappearing taxa, possibly contributing, over the long term, to the preservation of primate diversity.

Another time-scale expected to be an important factor for behavioral flexibility by adult females is the estrous or menstrual cycle. For adult female mantled howlers in two habitats of Costa Rican tropical dry forest, average cycle length has been estimated to be approximately 16 days, and female cycles are not synchronous, though they may overlap (Jones, 1985a). The latter condition may influence female–female competition for limiting resources, in particular, food, but also for males varying in quality. Demonstrating that individuals do not necessarily differ in their behavioral "decisions" this study shows that females exhibited sexual solicitations (the "rear-present" posture) to all adult males in the hierarchy equally, suggesting that all males are equally good mates. Other studies on these animals show, however, that males are subject to social parasitism by females (Jones, 1997a; see Chapters 3 and 6); thus, lack of discrimination by females may mean that all males are equally acceptable as partners to be parasitized and/or exploited rather than as partners for copulation. Field experiments are required to test these possibilities and to "unpack" the differential responsiveness of females to males as a function of female cycle stage for both related and unrelated individuals of all ages (see, for example, Jones, 1978, pp. 71–80).

Abiotic (e.g., weather) and biotic (e.g., plant phenology; social regime) conditions may also vary on very short time-scales (relative to T), possibly favoring behavioral flexibility. Rapid environmental (condition-dependent) changes (relative to T) may decrease the utility of a polyspecialist strategy or genetic polymorphism, possibly favoring a generalized response (e.g., personality; stereotyped and ritualized responses). Below some threshold value of environmental change, however, it might be beneficial for an individual to assess different (fitness) optima and to "track" changes with flexible behavioral responses. For an "energy-maximizer" (a female), changes in the dispersion and quality of limiting food convertible to offspring are expected to be of particular import for the organization and reorganization of action patterns (Silk, 1993; Jones, 1980, 2004). For example, female mantled howler monkeys experiencing food stress may be most likely to disperse from their natal groups (Zucker *et al.*, 2001; Jones, 1980).

Alloparental Behaviors as an Example of the Flexibility of Responses by Female Primates

Students of mammals have long recognized the significance of studying alloparental care (Spencer-Booth, 1970; Hrdy, 1976; Ross and MacLarnon, 2000) since the presence or absence of alloparental behavior may serve as a diagnostic criterion for higher grades of sociality (e.g., cooperative breeding or eusociality, see Chapters 8 and 9). Higher grades of sociality may be associated with greater behavioral flexibility where neural plasticity and number controls these responses, requiring greater brain development. Ongoing changes in social relations may be particularly salient for females who are expected to have greater confidence about coefficients of relatedness (r) compared to males in the same conditions and, thus, greater confidence in their overall strategies of assessment (Buchan *et al.*, 2003; Sherman and Neff, 2003).

After reviewing the literature on nonmaternal care in anthropoid primates, Ross and MacLarnon (2000) concluded that mothers may benefit from allocare in some circumstances but that these females are not likely to allow allocare if the costs (e.g., to infant survival) are high. Since females are expected to be energy limited, their decisions to permit or not to permit allocare of infants may depend, ultimately, upon the relationship between energy savings and expected future reproductive success. Studying mantled howler monkeys in Guanacaste, Costa Rica, Jones (1978) found that adult females exhibited a significant degree of condition-dependent variability in their propensities to express infant transfer ("aunting behavior") and "agonistic buffering." Infant transfer was first described for mantled howlers by Glander (1975). Characteristic of this behavior is the mother's tolerance of the proximity to and tactile exploration of her infant by other individuals, usually other females. The likelihood that "aunts" are the mother's kin has been pointed out (Crook, 1971; Wilson, 1975; Hrdy, 1976, 1999b), although the presence of female dispersal in the genus *Alouatta* makes it unlikely that coefficients of relatedness within groups are high (Jones, 1980).

Glander (1975) and Hrdy (1976, 1999b) proposed that infant transfer behavior represents an energetic benefit to the mother, a possibility consistent with the energetic strategy proposed by Schoener (1971) for females. Crook (1971) has also proposed that infant transfer behavior is beneficial to the infant in case of the mother's injury or death, or to be advantageous to the recipient of the infant who learns how to nurture an infant (Crook, 1971; Hrdy, 1976, 1999b). In a study of infant transfer in humans, it was proposed that the behavior affords the recipient of the infant the opportunity to exploit or to manipulate the phenotype of another's offspring to the recipient's reproductive and/or phenotypic advantage (Jones, 1986; see Chapter 3). Table 5.2 presents my observations of infant transfer in two groups of mantled howlers in two tropical dry forest habitats (riparian: group 5; deciduous: group 12), including the dominance ranks and sexes of individuals involved in the interactions (Jones, 1978).

"Agonistic buffering," first described in the barbary macaque (*M. sylvana*) by Deag and Crook (1971), entails a subordinate individual taking an infant from the body or proximity of its mother and used—presumably as a token of appeasement—to approach a dominant. These authors termed their observations "agonistic buffering" because the infant appeared to "buffer" the interaction between subordinate infant-holder and dominant target. Table 5.3 displays my observations of "agonistic buffering," including dominance rank and sex of individuals involved (Jones, 1978). "Agonistic buffering" was observed only in the riparian habitat group (group 5). For events of infant transfer and "agonistic buffering," lower-ranking individuals appear to be more likely to donate their offspring to higher-ranking individuals. While "agonistic buffering" in barbary macaques occurred only between subordinate and dominant males, female mantled howler monkeys were most likely

Table 5.2. Events of Infant Transfer from Mother (Donor) to Alloparent (Recipient) in Groups 5 and 12.

Mother	Alloparent
Group 5	
LL (5)	GWS (6)
YS (12)	GWS (6)
BC (13)	RS (14)
PY (16)	GS (7)
?	GWS (6)
LL (5)	GWS (6)
PY (16)	BC (13)
UM$_3$ (11)	GS (7)
YS (12)	OP (9)
PY (16)	PS (15)
UM$_3$(11)	RS (14)
BC (13)	LL (5)
BC (13)	RS (14)
PS (15)	R♂ (3)
PS (15)	GRS (4)
GRS (4)	GS (7)
GWS (6)	YS (12)
RYS (18)	GWS (6)
Group 12	
TC (6)	GS (5)
RS (9)	PS (3)
TC (6)	S♂ (1)
RS (9)	GS (5)
YPS (10)	S♂ (1)
RS (9)	GS (5)
RS (9)	GS (5)
RS (9)	S♂ (1)
TC (6)	PS (3)
RS (9)	GS (5)
RS (9)	GS (5)
RS (9)	GS (5)
TC (6)	GS (5)
YG (4)	RPS (7)
RS (9)	YG (4)
RS (9)	GS (5)
?	RS (9)
RS (9)	GS (5)

NB: Dominance rank in parenthesis: Jones (1978). All animals are adult females unless otherwise noted.

Table 5.3. Observations of "Agonistic Buffering."

Date	Mother	Recipient	Individual approached
8 February 8, 1976	PY (16)	GWS (6)	Y♂ (1)
February 22, 1976	RYS (18)	GRS (4)	R♂ (3)
February 23, 1976	?	GRS (4)	Y♂ (1)
February 23, 1976	?	GRS (4)	Y♂ (1)
March 8, 1976	LL (5)	MJ	G♂ (2)
March 9, 1976	LL (5)	UM$_1$ (8)	Y♂ (1)
March 17, 1976	LL (5)	LL (5)	GRS (4)
March 17, 1976	LL (5)	LL (5)	GS (7)
March 17, 1976	LL (5)	LL (5)	BC (13)
April 2, 1976	LL (5)	UM$_1$ (8)	R♂ (3)

NB: Group 5: dominance ranks in parenthesis, including Mother of Infant, Recipient of Infant, and Individual Approached by Recipient with Infant (Jones, 1978). MJ = medium juvenile. All individuals are adult females unless noted otherwise. Note that on three occasions, LL used her own infant to "buffer" interactions.

to approach dominant males with an infant "buffer" (six times out of ten). On three occasions (February 8, 1976, February 22, 1976, and March 9, 1976), the approaching female was in "peak" estrus and, presumably, ovulating. In one of these three cases (March 9, 1976), the male copulated while the infant was still on the female's body, supporting the view that "agonistic buffering" is conciliatory and that the infant was employed by the female to induce a resistant male to mate (see Jones, 1985a for documentation that males sometimes rejected females' solicitations to copulate). Thus, while the specific context of "agonistic buffering" seems to differ for barbary macaques and howler monkeys, the behavior patterns are similar and the function conciliatory in both species, suggesting convergent evolution of a flexible response sensitive to sexual (social) competition ("intrasexual" in the case of *M. sylvanus* and "intersexual" in the case of *A. palliata*).

Further support for the view that infant transfer and "agonistic buffering" by female mantled howlers represent reversible intraindividual behavioral flexibility is the finding (Tables 5.2 and 5.3) that the same individuals tend to be involved repeatedly in these interactions. Thus, Table 5.3 shows that LL's infant was utilized as a "buffer" in six out of ten events of "agonistic buffering", and GRS and LL use infant "buffers" six times out of ten events. Both of these females were high-ranking, holding first (GRS) and second (LL) positions in the female hierarchy of group 5. GRS was a young female who had not yet reproduced, and LL was multiparous. By January 1978, GRS had maintained her status, but LL had dropped to fourth position in the hierarchy. Two of the three recipients of LL's bouts of "agonistic buffering" were young, high-ranking females, including GRS, and GS was the fourth-ranked female in 1976 and 1977. By 1978, GS ranked second having assumed LL's position. Kinship is known for six of these events. MJ and UM$_1$ were female offspring of LL, and

LL used her own infant on three occasions as a "buffer." Thus, the possibility exists that an infant is related to the individual using it as a "buffer." These observations, also, highlight the hierarchical and social effects discussed by West-Eberhard (2003; also see Gross, 1996; Jones and Agoramoorthy, 2003) as part of an individual's assessment strategy for decision-making relative to the differential costs and benefits of social behavior ("reproductive" in the "West-Eberhardian/Alexandrian" sense). In the case described here, for example, it might be speculated that the frequent employment of "agonistic buffering" by GRS and LL occurred in the context of competitive relations for dominance rank, conditions expected to favor alternative behaviors.

The events displayed in Table 5.3 not only occurred in association with social changes (changes in the female hierarchy) but also with heterogeneity of food resources since all of these events occurred during three consecutive dry season months when preferred food (new leaves, flowers, and fruit) are available, distributed in a patchy manner (Jones, 1996b; Glander, 1975). It is possible that the events of "agonistic buffering" reflect social (reproductive) competition between individuals competing for ephemeral food. Supporting this view, nine of ten recipients of infants during encounters of "agonistic buffering" involved individuals who were to become or who were already dominant to the individual presenting the infant, suggesting that this condition-dependent response serves selfish individual interests.

Life History Tactics and the Evolution of Behavioral Flexibility

Observations of female mantled howler monkeys support the view that behavioral flexibility is competition-dependent (Crook, 1972; Gross, 1996; West-Eberhard, 1979, 2003; Jones and Agoramoorthy, 2003). Compared to mammalian males, mammalian females in the same conditions are expected to be particularly sensitive to costs in energy due to the expenses of parenting effort (offspring growth and development, lactation, and parental care; Wittenberger, 1980; Trivers, 1972; Shuster and Wade, 2003, Chapter 5). Errors in decision-making are expected to be very costly for mammalian females, a condition which will have important implications for the behavior of female primates and other female mammals, who are likely to be error and risk averse compared to males in the same conditions. If females adopt behavioral tactics and strategies to minimize risk, error, uncertainty, and unpredictability (see Chapter 9), knowledge of a population's life history strategy should provide insights into the behavioral responses favorable to most individuals in particular environmental regimes.

Table 5.4 displays an approximate life table for female mantled howler monkeys in tropical dry forest environment at Hacienda La Pacifica, Guanacaste, Costa Rica (Jones 1997b). A graphical representation of the

Table 5.4. Approximate Life Table for Female Howler Monkeys at Hacienda La Pacifica.

Age Interval	N	S_x	d_x	$1000q_x$	m_x	$l_x m_x/1000$	$x l_x m_x/1000$	v_x
I, 0–1 (0.5)	17	0.58	420.0	420	0.00	0.00	0.00	2.23
J, 1–4 (2.5)	16	1.22	0.0	0	0.00	0.00	0.00	4.46
SA, 4–5 (4.5)	12	1.54	0.0	30	1.00	0.58	2.61	4.46
D1, 5–7 (6.0)	37	0.97	17.4	660	2.00	1.13	6.78	3.30
D2, 7–10 (8.5)	54	0.34	371.3	910	2.50	0.48	4.08	2.61
D3, 10–15 (12.5)	31	0.09	174.1	950	2.50	0.04	0.48	2.59
D4, 15–25 (20.0)	4	0.05	16.3	1000	2.00	0.002	0.04	1.99
						2.23	13.99	

SUMS

$R_0 = 2.23$

$T = 13.99/2.23 = 6.27$

$r = (\log_e 2.23)/6.27 = .13$

$MLL = 5885.65/999.99 = 5.89$

NB: Jones, 1997b, used with permission. I = Infant; SA = Sub-adult; D = Adult age classes 1–4 (see Fig. 5.1)]. Numbers in the first parenthesis after each age class are those years spanned by the interval and numbers in the second parenthesis after each age class are the mid-points of each age class. The mid-point of each age class is employed as "x" in computations (see legend for Fig. 5.1). Survival (S_x) for each age interval was estimated from Scott's census (Malmgren, 1979) and defined as the ratio of numbers sampled at successive ages (after Smith and Polacheck, 1981). Estimates of survival from Scott's census allowed calculation of traits, assuming a cohort of 1000 newborns; number of individuals dying in successive age intervals (d_x); age-specific survival rate (l_x); age-specific mortality rate (q_x); fecundity (m_x), the average number of offspring each female will produce at age "x" (estimated from Clarke and Glander, 1984; Jones, unpublished); reproductive value (v_x), the relative number of offspring that will be produced by each female surviving till age "x"; net reproductive rate or generation size, the total expected reproduction by a female during her lifetime (R_0); the average age at which a female reproduces or generation time (T); the intrinsic rate of increase or exponential rate of population growth (r); and mean length of life (MLL). Across the S_x column, low survivorship appears to predominate during the infant and adult age classes. Smith and Polacheck (1981) point out that survival estimates >1 indicate high levels of variance in some mammalian studies. At La Pacifica, survivorship may be low or unpredictable throughout the period of immaturity. Equations after Odum (1971). See text for further explanation.

rate of decline in survival (l_x) for these animals (Fig. 5.4) yields a convex or "stairstep" function common among herbivores in high density conditions where survival rate is low or changes abruptly in more than one age interval (Taber and Dasmann, 1957; see Fig. 5.4). These patterns may be a result of different environmental or other constraints operating upon survival and reproduction at different times (Hamilton, 1966). The "stairstep" function is thought to be highly indicative of underlying density-dependent processes (Taber and Dasmann, 1957) such as resource depletion, interference, use of marginal habitat and resources, or predation. Further analysis of life history characteristics of these females (Jones, 1997b) demonstrated a relatively low rate of intrinsic increase, similar to other studies (Glander, 1980), and 1-, 6-, and 12-month rainfall intervals deviating significantly from unpredictability (temporal predictability; Jones, 1997b). Stearns (1976, 1992) showed that in these conditions adult survivorship will be favored over survivorship of infants and juveniles. Consistent with theory (Stearns, 1976; Geisel, 1976; Barclay and

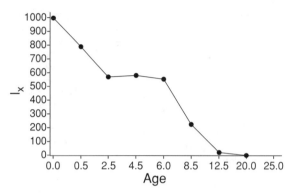

Fig. 5.4. Approximate survivorship curve (l_x) for female mantled howlers at Hacienda La Pacifica (Jones, 1997b).

Gregory, 1981; Partridge and Harvey, 1988), Jones (1997b) hypothesized that female mantled howlers were likely to practice a "bet-hedging" strategy, sampling the environment over time with offspring as currency in order to reduce variability in lifetime reproductive success. Mantled howler females demonstrate a suite of characteristics consistent with this view (e.g., iteroparity, relatively low reproductive effort). "Bet-hedging" will favor behavioral flexibility in females (where a tradeoff exists between reproductive effort and adult survival) because this strategy will lead females to sample a broad range of conditions in order to maximize fitness, yielding, perhaps, the responses described in this chapter as well as other flexible programs (e.g., flexible "mothering styles," Clarke, 1990; Nicolson, 1987; "maternal effects," Mousseau and Fox, 1998a,b; Watson *et al.*, 2003).

Infanticide by Females as a State-Dependent Flexible Response

Infanticide (or the threat of infanticide, Johnstone and Cant, 1999) is an alternative reproductive phenotype that is most likely to be perpetrated by males in heterogeneous regimes during changes in male tenure (van Schaik and Janson, 2000). Infanticide by females has also been observed (Digby, 2000). Andrews (1998), for example, reported one case of infanticide by a female black lemur (*Eulemur macaco*) in disturbed habitat. Reviewing the literature on infanticide in monogynous and cooperatively breeding callitrichids, Saltzman (2003) showed that infanticide is most likely to be perpetrated by females in the late stages of pregnancy, often associated with captive conditions or, in nature, with habitat disturbance or habitat fragmentation. Saltzman (2003, p. 213) makes the point that these crowded and/or heterogeneous conditions may "increase the likelihood of infanticide." Callitrichids should be used as a model for the descriptive and experimental investigation of infanticide by females in cooperatively breeding species because of the variability of infanticidal behavior as summarized by Saltzman (2003) and because of relatively clear patterns of this response such as the prevalence of breeding, dominant females killing the young of subordinate females, the apparent tendency for callitrichid females to commit infanticide in the late stages of pregnancy, the occurrence of both aggressive and affiliative behavior between breeding females in a group, and the potential for comparisons and contrasts between patterns of female infanticide among noncooperatively breeding and cooperatively breeding callitrichids as well as between cooperatively breeding callitrichids and other cooperatively breeding taxa, including the extent to which female infanticide is generally a condition-dependent response associated with heterogeneous regimes.

Kappeler (1999, p. 18) proposed that infanticide by males "has been a pervasive force in primate social evolution," and the topic of male infanticide as a sexually selected male reproductive tactic has received broad coverage in the primate literature (van Schaik and Janson, 2000). Digby (2000, p. 423), however, argued that infanticide by females "is likely to be taxonomically more widespread and, for group-living females, potentially a more constant threat than other forms of infanticide." According to Digby (2000; also see Hrdy, 1976), infanticide by females reflects female–female reproductive competition and is most likely to result from competition for resources (e.g., food) and competition for benefits to be derived from exploiting the young [e.g., intraspecific predation and "live use" (e.g., the "phenotypic manipulation" hypothesis; Jones, 1986)]. Digby's (2000) review shows that infanticide by females has been documented in several primate species; however, the most extensive studies of this phenomenon in primates have been conducted with the cooperatively breeding Neotropical callitrichids (marmosets and tamarins). Among these species, infanticide by females occurs almost exclusively by dominant females targeting the offspring of subordinate females (Digby, 2000; Saltzman, 2003).

Although Digby (2000) rejects sexual selection as an explanation for infanticide by females, all of the examples discussed in his chapter might be interpreted from the perspective of parental investment theory (Trivers, 1972) and a female's alternate reproductive tactics and strategies (see Jones and Agoramoorthy, 2003). If a female's allocation of time and energy are more likely to involve investment in parenting effort, then her responses toward other females as well as young might be explained as intrasexual selection whereby infanticide would be selected because it enhances mating success. In this view, infanticide is a form of post-copulatory intrasexual selection (see Alexander *et al.*, 1997). Infanticide by female callitrichids appears to conform to this model (see Saltzman, 2003). Finally, Crockett and Janson (2000) advance the hypothesis that infanticide by males limits group size in primates, an idea that might apply, as well, to infanticide by females where the act is likely to minimize within-group competition for limiting resources.

Conclusions

Despite these relatively straightforward predictions about the relationship between behavioral flexibility in females and environmental conditions, females are, nonetheless, expected to be more conservative than males in the same conditions because females will generally be energy limited. Although early work on the behavior genetics of males and females appeared to show that females are more "canalized" than males in the same conditions (e.g., Lerner, 1970), recent assessments of the behavior of female primates have stressed the flexibility and diversity of their repertoires (Hrdy, 1999a,b). Empirical studies are required to evaluate this apparent contradiction and the extent to which females may be selected, on average, to exhibit more conservative responses compared to males, *ceteris paribus*. Greater conservatism in female behavior, if demonstrated, may result from the greater vulnerability of this sex to social parasitism and other forms of risk (e.g., force or coercion) as well as greater canalization of female responses compared to those of males. Finally, Heinze and Keller (2000) proposed that higher grades of sociality may be favored for their benefits as energy-saving strategies. If this inference is a general principle, applying to social mammals as well as social insects, then typically higher expressions of social behavior and interaction rates among females, including primate females, may be explained by differential savings in energy. Studying cooperatively breeding callitrichids, Bales *et al.* (2000) concluded that "a reduction in energetic investment may translate into increased survival," a view in apparent support of Heinze and Keller's (2000) evolutionary scenario.

Male Primates: "Time-Minimizers" in Heterogeneous Regimes

6

In primates, male monopolization of female groups is thought to increase (resulting in fewer males in the group) when there are fewer females in the group (the spatial effect) and when females are less synchronously receptive (the temporal effect).

Nunn (1999, p. 1)

Introduction

As the sex with the lowest initial investment in reproduction (Trivers, 1972; Queller, 1997), the higher potential for lifetime reproductive success, and the higher variance in reproductive success, males are expected to be "time minimizers" (Schoener, 1971; Tolkamp *et al.*, 2002; Newton-Fisher, 2002), factors that are likely to influence decision-making by males in the face of environmental unpredictability, uncertainty, and/or risk. A major source of environmental uncertainty, unpredictability, error, and risk for males will depend upon their differential ability to discriminate their own offspring, their full-sibs, their fathers, and the mothers of their offspring, although males are expected to be able to discriminate their mothers and their mother's offspring with greater confidence (see West-Eberhard, 1975). These conditions generally will increase the difficulty for males in estimating r_{xy} and/or r_{xe} (e.g., Pound, 2002; see Chapter 2) compared to females in the same conditions (see Chapter 5). Neff (2003) recently found that male bluegill sunfish (*Lepomis macrochirus*), for example, display behavioral flexibility by adjusting their behavior toward young according to their confidence of paternity. In primates, Anderson (1992), studying Chacma baboons (*P. cynocephalus ursinus*),

79

provided evidence in support of the view that variations in the ability to estimate paternity as well as changing social conditions (e.g., availability of estrous females) were associated with paternal investment, and Buchan *et al.* (2003; also see West-Eberhard, 2003, pp. 289–291) recently showed that male baboons (*P. cynocephalus*) are able to discriminate those young they are most likely to have sired, extending support to these putative offspring during conflicts. Importantly, Anderson (1992) suggests that the propensity of baboon troops to subdivide, a phenomenon expected to reinforce association among relatives, may be an important factor in a male's ability to discriminate his young and/or to estimate his own or other males' likelihoods of paternity.

Uncertainty in the estimation of degrees of relatedness may also explain why males generally are higher risk-takers than females. It has been shown for damselflies (*Calopteryx maculata*; Waage, 1988), for example, that contests are more likely to escalate in the face of uncertainty. Primate males are expected to take greater risks for the same reasons that they are more likely than females to escalate interactions—the potential benefits are greater and a given reproductive loss, all other things being equal, costs less (see Parker, 1974; Trivers, 1972). If males are more likely to encounter heterogeneous conditions compared to females in the same regimes, males may have been selected to tolerate a higher level of complexity, on average, compared to females (see Sonsino and Mandelbaum, 2001).

The physical, cognitive, and other rigors imposed by uncertainty, however, may sometimes favor males who exhibit aggressive restraint (e.g., contests, persistence, persuasion; Jones, 1997a, 1996b; Jones and Agoramoorthy, 2003; East *et al.*, 2003) rather than force or coercion where differential advantages or disadvantages to genotype and/or phenotype are difficult or impossible to assess. This perspective supports the view that behavioral flexibility occurs as a result of assessment in different situations relative to variations in potential benefits and costs (Parker, 1974; Enquist and Leimar, 1983; Eberhard, 2003, pp. 449–457). Figure 6.1, for example, displays alternative reproductive behaviors for male black howler monkeys (*Alouatta pigra*), identifying the array of responses that males might adopt over time and space and demonstrating manifestations of flexible behavioral responses.

Errors in decision-making will, all other things being equal, impact a male's inclusive fitness less than a female's in the same conditions because of anisogamy (small initial investment in reproduction; Trivers, 1972), because males are not as responsive to energy limiting effects (up to some threshold value) relative to females in the same regime (Schoener, 1971), and because investment in parenting effort is, on whole, less energetically costly for males, particularly for male mammals (Queller, 1992). The foregoing advantages for the male sex may be outweighed by potential disadvantages (e.g., reduced survival), however, because increased phenotypic plasticity in males will be correlated with extreme phenotypes (e.g., Relyea, 2002; Alexander *et al.*, 1997), because females are a limiting resource for males (Trivers, 1972), and because

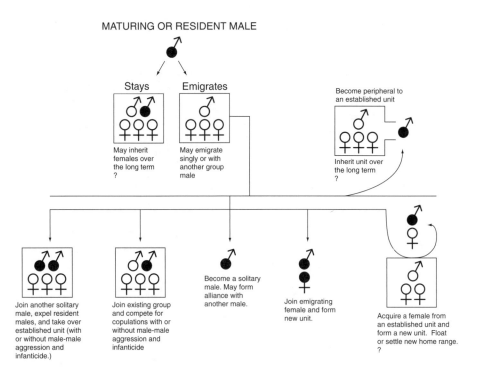

Fig. 6.1. Alternative reproductive behaviors identified for male black howler monkeys (*A. pigra*) representing the array of responses that might be adopted for inclusive fitness maximizing (after Horwich *et al.*, 2000). The possible responses include remaining in the natal group to breed, joining an established group, and colonization. Behavioral flexibility implies that individual males may adopt more than one of these responses during his lifetime.

variance in reproductive success among males competing for the same females will often be high (see Trivers, 1972). Where these potential costs are deleterious to inclusive fitness, say, because of high assessment costs (error; see Luttbeg, 2004), low population density, female "choosiness," temporal or spatial unpredictability of limiting resources and the reproductive females who utilize them (leading to female "emancipation"; Jones and Cortés-Ortiz, 1998; Emlen and Oring, 1977), or because of differences in male quality (e.g., as a result of age, size, fighting ability, nutritional state, or fatigue); males may display behavioral flexibility to minimize costs and maximize benefits to inclusive fitness. In these circumstances, males may adopt alternative tactics or strategies such as policing (Jenkins *et al.*, 2000), interference or disruption (e.g., sperm plugs; see Dixson, 1998), persistence (e.g., queuing; Alberts *et al.*, 2003), persuasion (e.g., courtship; Jones, 1995a, 1996b, 1997a), or social parasitism (e.g., "sneaking," dispersal, exploitation, mimicry; Jones and Agoramoorthy, 2003; Double and Cockburn, 2003; Shine *et al.*, 2001; Taborsky, 1994).

The "Branch-Break" Display of Male Mantled Howler Monkeys

An example from mantled howlers highlights the importance of condition-dependent alternative responses to males. Male mantled howler monkeys (*A. palliata*) display a variety of stereotyped and ritualized behaviors that appear to be facultative responses to situational changes (Jones, 2002a, 2003a). These responses have been interpreted as condition-dependent displays of quality, and the "branch-break" display reflects contest competition in which two or more individuals encounter each other and compete for a limiting resource (e.g., females). In these conditions, individuals may resolve the conflict either by fighting until one individual wins the contest or by exhibiting displays for the assessment of relative strength between competitors, avoiding the risks of serious injury. Displays are likely to evolve where the costs of fights are high (Krebs and Davies, 1993) and are considered to represent stereotyped or ritualized intention movements, ambivalent responses, or redirected acts (Tinbergen, 1952). The following discussion of branch-break displays exemplifies how behavioral flexibility may be employed in response to local competition to coordinate and, possibly, to control interindividual conflicts of interest as well as to optimize inclusive fitness.

Branch-break displays in mantled howlers generally entail an adult male initially breaking a branch from a tree, moving it with his arm(s) horizontally and/or vertically with varying degrees of force (intensity) at one or more adult males in what appears to the human observer to be an exaggerated and intentional visual performance. The "arch display," a humped-back posture (Crockett and Eisenberg, 1987, p. 56; Carpenter, 1934, p. 101), was sometimes incorporated into the action patterns of the branch-break display. Individuals either remained in one position in a tree or, more commonly, walked or ran sideways or back and forth along the limb of a tree during the display or, in some instances, pumped up and down or sideways while swaying the branch being held. Because the sender or the receiver of these action patterns is always displaced, branch-break displays were considered to be displays of threat. The behavioral ecology literature, however, generally assumes that male threat displays may also function as displays of quality to potential mates (Krebs and Davies, 1993; Andersson, 1994; Luttbeg, 2004).

As Bradbury and Vehrencamp (1998, p. 598) pointed out, "threats occur by definition in situations of sender–receiver conflict," favoring a close functional association between the form and content of threat signals. Carpenter (1934) reported threat displays by male mantled howlers entailing vigorous branch shaking, combined with various bodily postures, excited movements, and vocalizations. Branch-break displays in this species were documented anecdotally by Glander (1975) who reported observing them most commonly during intergroup interactions. Jones (2000) presented empirical evidence to show that branch-break displays were the most frequent form of agonistic action patterns displayed by adult males in her riparian study group in Costa Rica. Threat displays incorporating branches have been observed in other primate species, including *Pan troglodytes* (van Lawick-Goodall, 1968), *Hylobates lar* (Ellefson, 1968), *Macaca fuscata* (Wolfe, 1981; Soltis *et al.*, 1999 and references therein), *M. fascicularis* (Palombit, 1992), *Cebus capucinus* (Oppenheimer, 1973), and *Ateles geoffroyi* (R. Horwich, personal communication). The apparently purposeful breaking of branches for use in displays, as occurs in the branch-break display of mantled howlers, has rarely been described in the primate literature. In this chapter I present quantitative data on branch-break displays by adult male mantled howlers, a flexible and stereotyped behavioral pattern that appears to express conditional signals of threat in the context of competition with rivals for mates. These displays may also function to attract females, and branch-break displays may be components of "compound displays" which enhance their signal function and versatility. By analyzing the results presented here in relation to signaling theory, examples are provided of how local competition may favor the expression of behavioral flexibility.

Investigating Behavioral Flexibility in Male Mantled Howler Monkeys: Study Sites, Procedures, and Definitions

This study (Jones, 1980, 1985a, 2000 and references therein) was conducted in 1976 and 1977 at Hacienda La Pacífica, Cañas, Guanacaste, Costa Rica ($10°18'$ N, $85°07'$ W). Modal social organization of mantled howlers is multimale–multifemale, yielding a polygynandrous mating system (Crockett and Eisenberg, 1987; Carpenter, 1934; Jones, 1980, 1985a, 1995a, 1998, 2000; Glander, 1980). Two marked groups were studied in two habitats of seasonal, tropical dry forest (riparian habitat, group 5: three adult males plus a subadult male, LT, 15 adult females, 402 h observation; deciduous habitat, group 12: two adult males, eight adult females, 114 h observation). In group 5, Y male was highest ranking, G male, second ranking, and R male, lowest ranking (see Jones, 1980 for procedures used to determine dominance rank in both groups). LT male had not secured group membership or a position in the male hierarchy at the time of the present study. After the present study was concluded, the young LT male entered the hierarchy in 1977 as dominant male resulting from a coalition with Y who fell to second rank. G was expelled from the group as a result of this coalition, and R remained lowest ranking (Jones, 1980). In group 12, S male was dominant, Z male, subordinate. In previous reports (e.g., Jones, 1980), Z male was labeled "R_{12}". To avoid confusion with R male of group 5, this group 12 male has been relabeled Z. Results are based upon randomized focal (Altmann, 1974) and *ad lib.* observations (Jones, 1978).

I use "display" to mean one or more branch-break events (BBE) in a 30-min interval where an interval begins with the initiation of a BBE. A BBE was determined to begin when a male (the sender) selected a branch and initiated the behaviors described above, always in the direction of another male (the receiver). The receiver of a display was determined to be the male who was the nearest neighbor of the displaying male, always within 20 m. Two males were presumed to be the receivers of a display if the two were within 20 m of the displayer and were judged to be in coalition against the displayer (e.g., by jointly displacing the displayer within 1 h of the focal BBE). Termination of a display (one or more BBEs per interval) was defined as termination of the behaviors described above and displacement (i.e., one male, either sender or receiver, moved at least 1 m away). Since focal observations were a minute-by-minute record of the focal animal's responses, each BBE is effectively equivalent to one minute of observation. A single BBE always implies one individual utilizing one branch, although multiple BBEs during a 30-min interval may imply that one individual has selected more than one branch with which to display. Branch-break displays may be repeated by the same male displaying singly or by an exchange of displays between two males.

All counts of behavioral events are reported with the exception of "vocalizations," which are reported as categories of behavior. Vocalizations vary in frequency, intensity, duration, and pitch, and some of these are expressed repeatedly at a high rate (e.g., "gutteral barks"; Jones, 2000; Baldwin and Baldwin, 1976), events that were not captured in raw data form. "Mount" indicates a dorso-ventral posture without intromission, and "copulation" means mounting with intromission, with or without ejaculation. "Approach" and "avoid" are defined as animals move 1 m toward or away from another, respectively. The nonparametric chi square "goodness of fit" test (two-tailed) and the normal approximation to the binomial (z-test) are employed for statistical analyses with alpha set at 5%.

How Does the "Branch-Break" Display Demonstrate Behavioral Flexibility?

Table 6.1 presents branch-break displays by adult males (N = 65) observed in groups 5 and 12, analyzed by signaler and receiver(s), total number of BBEs, and maximum number of BBEs per 30-min interval. Because of small sample sizes, one instance of multiple displays exchanged among adult males during an intergroup encounter (groups 5 and 7; see Glander, 1975) and four displays exhibited by adult females of group 5 are excluded from analysis. In these cases, branch-break displays appeared to function as for males—to displace another individual or group from a contested resource (e.g., mates, food, space).

Sixty-three BBEs were observed in group 5 (z-test, P = 0.0102), two in group 12 (z-test, P = 0.9898). Thus, BBEs in the riparian habitat occurred more frequently than would be expected by chance alone, while BBEs in the deciduous habitat did not. Hourly rates in the two habitats were 0.16 (riparian) and 0.02 (deciduous). Branch-break displays occurred in association with vocalizations (N = 31), including gutteral barks, howls, and paedomorphic whines (Jones, 1980), piloerection (N = 1), genital display (N = 1; Jones, 1985, p. 132), the "urine-wash" display (N = 1; Jones, 2003a), and/or a variety of bodily orientations (front-, side-, and rear-present postures; Jones, 2000), responses that may facilitate assessment of relative resource-holding potential by rivals and that may be components of a "compound display" (Bradbury and Vehrenamp, 1998). Compound displays are features of many animal communication systems (Bradbury and Vehrencamp, 1998) and would be expected to enhance the flexibility and complexity of any message emitted by a sender.

The three group 5 males, plus the sub-adult male, LT, differed significantly in total number of BBEs (Y male, 28 BBEs; G male, 6; R male, 28; and LT male, 1: $\chi^2 = 38.91$, df = 3, P $\leq$ 0.001). G and LT males were significantly less likely to display than Y and R. No significant difference in total number of BBEs occurred between the two males of group 12 ($\chi^2 = 1.00$, df = 1, P > 0.05). Comparing the total number of BBEs across all observed signaler → receiver combinations (Table 6.1, column 2), significant differences were found ($\chi^2 = 93.13$, df = 11, P $\leq$ 0.001). Group 5's R male was much more likely to display to that group's Y male than any other combination. Three displays escalated to chases or fights, all involving Y and R males, and mutual exchanges of displays occurred twice, each time involving R male as the initiator to Y. The modal number of BBEs per sequence was 1 BBE in 1 min (n = 24), and only Y and R males exceeded this limit (Table 6.1, column 3).

Table 6.1. Direction of "Branch-Break" Displays (Male Sender → Male Receiver[s]), Number of Events Observed, and Maximum Number of Events per 30-min Interval.

Direction of display (signaler → male receiver[s])	Number of "branch-break" events (BBEs)	Maximum number of BBEs/interval
Y → G	7	5 BBEs in 19 min
Y → R	11	3 BBEs in 23 min
Y → G & R	9	2 BBEs in 19 min
G → Y	3	1 BBE in 1 min
G → R	1	1 BBE in 1 min
G → Y & R	2	1 BBE in 1 min
R → Y	24	4 BBEs in 9 min
R → G	3	2 BBEs in 2 min
R → Y & G	1	1 BBE in 1 min
LT → G	1	1 BBE in 1 min
Z → S	2	1 BBE in 1 min

This finding supports the view that repetition of displays is energetically costly (Payne and Pagel, 1997), consistent with Strier's (1992, p. 515) conclusion that howling monkeys exhibit a behavioral repertoire "in which energy expenditure is minimized." These results suggest that decisions to employ flexible signals and displays require prudence on the part of the sender and receiver.

All displays occurred in the presence (i.e., within 20 m) of one or more cycling females when one male (i.e., sender or receiver) and/or a cycling female were judged to exhibit sexual behavior (e.g., consort behavior or mate-guarding by males, the "lingual-display," or sexual solicitations by either sex; Jones, 1985a, 1995a). Thus, branch-break displays may provide information to males for the assessment of a rival's resource-holding potential and to females for the assessment of mate quality (Andersson, 1994; Payne and Pagel, 1996a). On five occasions, displays were exhibited shortly before or after copulation (range = 1 min – 1 h 6 min) by Y (n = 3), G (n = 1), or R (n = 1) males. While a cycling female was sometimes observed to approach a male after he displayed, females appeared to be significantly "emancipated" from male control (Jones, 1985a, 1995a, 2000; Jones and Cortés, 1998; Emlen and Oring, 1977; see Chapter 1) since females approached more than one male and also avoided and/or rejected mounts and copulations before or after BBEs.

The "Branch-Break" Display by Male Mantled Howlers as a Compound Display of Threat

Compound displays are evidence of behavioral flexibility and the potential to switch from one response to another. The results presented here suggest that branch-break displays may be elements of a "compound display" (Bradbury and Vehrencamp, 1998) involving visual, auditory, and olfactory signals. Apparent enhancement of visual signals has been observed in association with threat displays in other animals (Bradbury and Vehrencamp, 1998). Although communication in howlers is "primarily vocal" (Carpenter, 1934; Chivers, 1969; Wilson, 1975; Baldwin and Baldwin, 1976; Jones, 2000), the present data show how visual signals can enhance auditory signals in this genus and contribute to the broad range of alternate responses characteristic of *Alouatta* species. Condition-dependent signals are a function of phenotype or environment ("best of a bad situation" rules; Brockmann, 2001) and are likely to be displayed when individual quality varies over time, possibly as a result of nutritional state or fatigue, factors that are not immediately visible to rivals (Payne and Pagel, 1996a). This variation is a component of the biotic environment's heterogeneity ("disturbance regime") whose assessment by individuals may minimize uncertainty, unpredictability, error, and/or risk (see West-Eberhard, 2003).

Signaling Theory and Patterns of Branch-Breaking in Mantled Howler Monkeys

The pattern of BBE repetition and frequency presented above can be interpreted in terms of theoretical models, bearing in mind that the overall "signal" magnitude may depend on combining and recombining a number of display elements (i.e., flexible behavior). The problem of understanding how the receiver combines information from a sequence of BBEs in order to make an assessment of the signaler is essentially a separate issue from understanding how the signaler decides what magnitude of signal to give. Both of these topics deserve investigation in primates.

Evolutionary models show that the form of a display (in terms of number of repetitions, or changes in intensity over time within a display sequence) is predicted to depend on the way the display is assessed and interpreted by the receiver ("assessment rules"; Payne and Pagel, 1997; Parker, 1974; West-Eberhard, 2003). The present results do not contain information on "intensity" in each BBE, but they do reveal how mean numbers of repetitions differ among individuals. The modal number of BBEs per sequence is 1 BBE/1 min, and BBEs per interval never exceeded 5 (Table 6.1, column 3). This low modal number of BBEs per sequence is not consistent with the predictions of the sequential assessment model (Brockmann, 2001) in which the "signal" is the average of the display elements so far, but it is consistent with an extreme case of a signal of endurance (Payne and Pagel, 1996a), in which the signal is the cumulative sum of the elements, accounting for any variation in intensity. The findings are also consistent with the "best-so-far" model (Payne and Pagel, 1996b), in which the signal is the intensity of the most recent element only. As Payne and Pagel's "best-so-far" model is summarized by Ord and Evans (2003, p. 1504), "When signals incur significant costs (e.g., from fatigue), assessment is more likely to be based on a cumulative measure of all displays performed." These authors continue, the model "is a cumulative function rule, together with a threshold that varies according to individual condition." The "best-so-far" model is unusual in that it is the only model that necessarily predicts that the majority of signal sequences will only contain one element, as is observed in the present data set.

Further observations of factors, such as any variation in element intensity, will be needed to resolve between the latter two models for the displays of mantled howler monkeys, and it will be necessary to systematically study the branch-break display in additional groups before confident statements can be made about its function(s). Nonetheless, it seems reasonable to assume, tentatively, that the signal magnitude is at least in some way proportional to the number of BBE repetitions and that these quantities, taken together, can be employed as an effective measure of behavioral flexibility. It is important for future research to investigate whether survival and/or reproductive and phenotypic success change as a function of changes in signal magnitude relative to the number of BBE repetitions and to evaluate all of the models discussed by Ord and Evans (2003) and Luttbeg (2004) as potential explanatory schemas for primate signaling systems.

There is something unusual in the present data. Discussions of indicator signals (sometimes called "honest" signaling or the "handicap principle") usually presume that better quality, or higher-ranking males give larger signals. The idea is that only individuals in good condition can afford costly signals, which thereby ensures signal "honesty" (Andersson, 1994). But, that is not the case here. Limiting analysis to males who had attained a position in the hierarchy, the middle-ranking male, G, displayed with low frequency compared to the highest and lowest ranking males, Y and R, and R was most likely to display to Y. The data for Y male are consistent with the handicap principle, but those for R male are not. Here it is not the highest-ranking males that signal most, but both the highest and the lowest (with the middle-ranking male signaling least).

At first sight this pattern of results appears to contravene theory of indicator signals. But for indicator signals to be evolutionary stable it requires that either the relative costs and/or the benefits must be different for some signalers than for others. The larger signals are given by those with differentially cheaper costs and/or by those with differentially greater potential benefits. Discussions of "honest" signaling usually only consider differential signaling costs, and therefore permit only that better quality males signal more (display more flexible behavior?). Yet, once differential benefits are considered, it is possible to think of a number of scenarios under which it is the weaker male that signals more—if weaker or lower-ranking males have relatively more to gain then they will give the larger signals (akin to "begging theory"; Godfray and Johnstone, 2000). By considering costs and benefits together it is, in principle, possible to have scenarios in which the larger signals are given by the highest and the lowest ranking males (as in the present case), or even scenarios in which the middle-ranking male gives the largest signals (R.J.H. Payne, personal communication; see also Proulx et al., 2002). We therefore need to understand the relative costs and benefits to the male howlers.

"Time-Mimimizing," Age, Rank, and the Use of "Branch-Break" Displays

Previous reports on this group have shown that dominance rank was positively and significantly related to frequency of copulation (Jones, 1985a, 1995a) and that G and R males were most likely to be in coalition against Y (Jones, 1980, 1982a, 2000). Thus, Y male may have been R male's primary constraint in the pursuit of reproductive success and his primary rival (Jones, 1982a, 1985a). Alternatively, R male may have concentrated his displays upon Y because, as the male most successful in attracting females and obtaining copulations, particularly with females in "peak" cycling stage (Jones, 1985a),

Y may have valued a single resource (i.e., a cycling female) less than other males, increasing R's likelihood of mating. In this case, costs (e.g., search costs, exposure to predation) would be, effectively, lower for Y. The assumption that R male has more to gain than Y or G has to lose may explain his investment in branch-break displays. This analysis highlights the importance of assessing the differential costs and benefits to males of variable quality for an understanding of behavioral flexibility (in this case, different patterns of branch-break displays, including compound displays).

It is difficult to be certain how the signal size, say, S, depends on male quality, say, Q, requiring, as it does, resolution of the issue of assessment rules mentioned above. But the frequency with which Y and R males exhibit and repeat branch-break displays (Table 6.1 and above discussion) can be employed as an indirect measure of S. It is also hard to tell whether dominance rank acts better as a surrogate measure of male quality or of resource value (since higher-ranking males get more access to cycling females). It is also important to note that, while females were known to be cycling, the precise cycling stage is unknown for most events reported here. Thus, resource value, V, has not been measured precisely.

For the present data, the lowest ranking male signals the most, and this implies that the value of winning is worth more to the lowest ranking male, slightly less to the highest ranking male, and least of all to the middle-ranking male. The middle-ranking male might employ strategies other than the "branch-break" display to obtain access to cycling females, such as coalitions with the lowest ranking male against the highest ranking male (Jones, 1982a, 1995a). Under this interpretation, *A. palliata* males who have attained group membership (Jones, 1980) appear to combine a condition-dependent strategy based upon predictions of the "handicap" principle for the highest ranking male (Y's use of the branch-break display) and, for the lowest ranking male, R, the prediction that lower-quality individuals have more to gain than the rival has to lose (Jones, 1980, p. 400; Parker, 1974). These combined results, whereby these two strategies are employed concurrently by males differing in quality, and the present interpretations, are to the best of my knowledge novel ones in the animal literature and, if confirmed, support the view that howlers, for whom rank is age-dependent (Jones, 1980), exhibit a significant degree of behavioral flexibility (Crockett and Eisenberg, 1987; Jones, 1995b; Crockett, 1998; Strier, 2000).

For hierarchical groups, the ranking introduces an extra factor that is not considered in most indicator models. In a hierarchical system, the simplest possible model could be that the cost-benefit outcome of an encounter is determined solely by the change in rank incurred so that the signal magnitude is determined not by the absolute male quality or rank (as most signal models assume) but by the rank difference of contestants. Thus, a top-ranking male may potentially suffer the greatest costs if it drops in rank, and a bottom-ranking male may potentially gain the greatest benefits if it rises in rank.

Under this scheme, it is natural that we expect the interactions between top- and bottom-ranking males to yield the most BBEs, and those interactions involving middle-ranking males to have the least BBEs. Although this provides an intuitive and easy explanation, the present data do not have enough resolution to make a significant test of whether relative rank is significantly better correlated to BBE repetition than is absolute rank for mantled howler monkeys.

It would be interesting to compare data from other hierarchical species since the present study highlights the significance of studying male tactics, strategies, tradeoffs, and interests differentially by quality (e.g., rank, age) and raises the intriguing question of whether the present pattern of results is explained best as a function of individual rank or rank difference between males. One possible method to measure rank difference would be to assess the mating success of males relative to one another. For the present group, G male's copulation success was more similar in quantity and quality (i.e., cycling stage of female partners) to Y than to R (Jones, 1985a, 1995a; see Chapter 4). These results suggest that the temporal scale of environmental heterogeneity is very important for determining whether or not patterns of response to local conditions reflect longer-term patterns of response by and interactions among individuals. Patterns of response as a function of male rank for two other displays exhibited by male mantled howler monkeys (the "genital display," Jones, 2002b; the "urine-wash display," Jones, 2003a), for example, differed from each other and from that described in this chapter for branch-break displays.

Further research needs to assess the reliability of the present results and to document the use of displays differentially by male rank (Jones, 2002c, 2003a). It will also be essential to measure the importance of branch-break displays in the attainment of group membership and rank by sub-adult males attempting to gain group membership (e.g., LT) and a position in a group's hierarchy. This display, for example, may be one employed only after a male has attained group membership and rank. Studies are also needed to describe in greater detail the alternative responses of middle-ranking males (e.g., G) relative to the highest ranking and lowest ranking males, in particular, the value for reproductive success of coalitions between middle-ranking and lower-ranking males (Jones, 1980, 1982a, 2000). The discussion in Chapter 4, for example, might lead to the speculation that G would employ cognitive mechanisms of categorization in his relations with Y, but mechanisms of discrimination in his relations with R. Finally, it would also be of interest to evaluate the possibility that the lowest ranking male (R) is "bluffing," comparable to the behavior of small, and presumably subordinate, male green frogs (*Rana clamitans*; Bee *et al.*, 2000). However, because of the theorized high costs of display to the lowest ranking male, this possibility seems unlikely. Furthermore, a previous report (Jones, 1995a) showed that R was the only male observed to fight another male (Y) to obtain access to an estrous female. Signaling a predisposition to escalate, then, may characterize low-ranking rather than high-ranking males in this species, with generally low rates of damaging aggression and with a broad repertoire of stereotyped and ritualized signals and displays among males (see Jones, 2000).

The Influence of Females on Male Tactics and Strategies

A mantled howler male's reproductive success is not only constrained by other males but also by females who are expected to prefer males with the highest mean fitness *among the males available to them*. The environmental potential for female choice will vary, and a high potential for female discrimination ("female emancipation") is thought to occur where unpredictable dispersion of resources utilized by females renders monopolization by males difficult or impossible (Jones, 1985a; Jones and Cortés-Ortiz, 1998; Emlen and Oring, 1977). Female emancipation has important implications for the analysis of male contest competition in *A. palliata* since it implies that males can never truly "own" the resource (i.e., the cycling female), increasing uncertainty and, thus, potential error (and risk) among males about relative resource-holding potential. Uncertainty among males about relative resource-holding potential will also increase the chances of escalation (see, for example, Waage, 1988). Both Carpenter (1934) and Jones (2000) concluded that stereotyped and ritualized responses were more common among male than among female *A. palliata*. If the dispersion of female mantled howlers creates a spatial and/or temporal configuration favoring male–male agonism, the broad array of stereotyped and ritualized action patterns, including vocalizations, documented for males of this species may have been favored by selection in response to pressures to minimize the costs of damaging aggression. This argument may apply, as well, to other primate species exhibiting low rates of direct aggression among adult males (e.g., *Brachyteles arachnoides*; Strier, 2000; Strier *et al.*, 2002).

Finally, an extension of Emlen and Oring's (1997) discussion of "female emancipation" is the concept of free female mate choice whereby viability of offspring is presumed to be greatest where females choose mates at random (Drickamer *et al.*, 2000; Gowaty, 1997). To date, the demonstration of this phenomenon has been limited to laboratory conditions that control for factors other than female mate preferences. It seems unlikely that, in nature, females of most species will be "free" from significant constraints upon their mating decisions, including taxa in which the sexes release their gametes "freely" into the environment (e.g., "free spawners"; Marshall *et al.*, 2004). Nonetheless, the concept of "free mate choice" and tests of it probably have the potential to reveal important insights into the proximate and ultimate causes and consequences of "female emancipation."

The pattern of results presented in this chapter suggests that male mantled howlers use branch-break displays as reliable and condition-dependent signals of threat in the context of reproductive competition, exhibiting behavioral flexibility as a function of age, dominance rank, and, possibly, variations in competitive regime and motivation over time and space. Consistent with the conclusions of Proulx *et al.* (2002) and the observations for mantled howler males reported in this chapter, female preferences may be influenced by the tendency for older males to exhibit more reliable and informative displays. Supporting this view, Luttbeg's (2004; also see Johnstone and Earn, 1999) theoretical treatment of the relationship between alternative male mating tactics and female mate assessment and choice shows that female behavior is strongly influenced by the ability to accurately assess male quality.

Conclusions

Future research on howlers and other primate taxa should weigh the likelihood that displays by males are utilized for the communication and assessment of information to resolve conflicts about who might be the fiercest opponent or the best choice of mate as a function of age, physical condition, relative dominance rank, etc. For male mantled howler monkeys, and, perhaps, for males of many other mammalian species, local competition for mates may favor the expression of behavioral flexibility for sending and receiving situationally optimal information. Such flexibility may require that individuals make decisions about the optimal organization of the reversible components of their phenotypes, including, in some instances, when to "switch" from one tactic or strategy to another. Higher cognitive processes, however, are not required for the expression of these and other flexible responses as demonstrated by their occurrence in insects and other taxa lacking complex neural networks comparable to those of mammals (see, for example, West-Eberhard, 1979). It seems a truism that animals are not required to *be* intelligent but that it is often beneficial for them to *act* as if they were. Chapter 7 addresses intersexual interactions, in particular, the inherent conflicts of interest between the sexes that, it is proposed, may lead to "antagonistic" responses between them which, in some conditions, may lead to an evolutionary "chase."

Intersexual Interactions in Heterogeneous Regimes: Potential Effects of Antagonistic Coevolution in Primate Groups

7

> Conflicts of interest between the sexes are manifest at all levels from behaviour to molecules, and such conflicts are key to our understanding of reproductive biology.
>
> Gavrilets *et al.* (2001, p. 537)

Introduction

If individuals differ genetically and in their initial investments in reproduction, genetic conflicts of interest may favor responses (e.g., male coercion of females; cryptic female choice) to optimize an individual's tradeoff of costs and benefits (to genotype and/or phenotype) relative to given abiotic or biotic (including social) conditions. Genomic conflict can occur in several ways (*intragenomic conflict, intergenomic conflict, intersexual ontogenetic conflict*; e.g., Pomiankowski, 1999; Rice and Holland, 1997; Rice and Chippendale, 2001; Chapman *et al.*, 2003 a,b; Hager and Johnstone, 2003). *Intergenomic conflict*, the focus of the present chapter, occurs when different optima exist for a trait expressed in different individuals (e.g., males and females), leading to a conflict between individuals over the most favorable optimum for the trait (see Pomiankowski, 1999; Hager, 2003c; Hager and Johnstone, 2003). It is thought that this form of conflict may occur with or without genetic correlation

93

(Chapman *et al.*, 2003a,b). Models of genomic conflict between the sexes are generally discussed as alternatives to models of sexual selection (male enticement and female attraction), in particular, indirect genetic benefits (Fisher's "runaway process" or "good genes models"; see Shahnoor and Jones, 2003) to explain interactions between the sexes and the behavioral and other features of males and females associated with these interactions. Theoretical work by Gavrilets *et al.* (2001; also see Gavrilets, 2000) and empirical studies by Moore *et al.* (2001), however, suggest that sexual conflict (male enticement and female resistance) may underlie the evolution of mechanisms of sexual selection. If this is the case, or if it is shown that sexual selection and sexual conflict are different processes, it is unlikely that sexual traits can be predicted in a straightforward manner from the conditions of anisogamy (Pagel, 2003).

Sexually antagonistic coevolution results from intergenomic conflict whereby individuals of each sex are favored to optimize their own tradeoffs of (genetic) benefits to costs (Parker, 1979). In primates, this process and its associated mechanisms may lead to behaviors such as infanticide (van Schaik and Janson, 2000; Crockett, 2003; Palombit, 2003; Saltzman, 2003), search strategies of mates by males (Fukuda, 2004; Jones and Agoramoorthy, 2003), high copulation rates (van Lawick-Goodall, 1968; Dixson, 1998), rape (Smuts and Smuts, 1993; Jones, 2002b), multiple mating by females leading to sperm competition (Dixson, 1998; Reeder, 2003; Harcourt, 1998; Jones and Cortés-Ortiz, 1998), "parentally biased favoritism" for particular offspring (Lessels, 2002), or social parasitism, including phenotypic manipulation, of one sex by the other (Jones, 1997a). While the investigation of sexual conflict is in its early stages in primates and other taxa (Rice, 2000; Nunn, 2003), in some species it has been demonstrated to influence components of fitness (e.g., zebra finches; Royle *et al.*, 2002).

The present chapter, intended to complement Chapter 8, assesses some possible outcomes of antagonistic coevolution in primates in relation to variations in environmental heterogeneity. Although Chapman *et al.* (2003a, p. 41) point out that "the boundary, if there is one, between traditional models of sexual selection and those of sexual conflict has not yet been carefully explored theoretically," I propose that antagonistic coevolution and sexual selection might be viewed along a continuum from relatively low to relatively high intensity (Fig. 7.1). This schema corresponds with relatively low to relatively high environmental heterogeneity, particularly food (see Nunn, 2003; Sterck *et al.*, 1997; Wrangham, 1979, 1980; Jones, 1981).

The positive relationship between sexual selection and environmental heterogeneity is well established (Emlen and Oring, 1977; Carlsbeek *et al.*, 2002), and I propose that the same relationship occurs between sexual conflict and environmental heterogeneity. These associations are expected, in turn, to be related to differential likelihoods of alternative sociosexual organization of individuals in time and space (Emlen and Oring, 1977; Bradbury and Vehrencamp, 1977; Shuster and Wade, 2003). Rice (2000) has pointed out

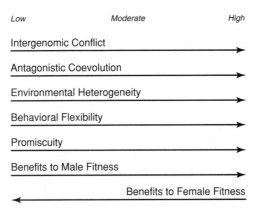

Low   *Moderate*   *High*

Intergenomic Conflict

Antagonistic Coevolution

Environmental Heterogeneity

Behavioral Flexibility

Promiscuity

Benefits to Male Fitness

Benefits to Female Fitness

Fig. 7.1. The sexual conflict continuum (SCC) from low to moderate to high intensities of sexual conflict. Sexual conflict implies that traits enhancing the reproductive success of one sex are costly to the fitness of the opposite sex (Gavrilets, 2000). To date, intensity of sexual conflict has not been determined quantitatively for any species. This graphic also shows that intergenomic conflict and antagonistic coevolution, environmental heterogeneity, behavioral flexibility, and promiscuity are all positively associated with an increase in sexual conflict. As shown by Rice (2000, p. 12953, Figure 1), the correlation for fitness of males and females in the same conditions are reciprocally related since, all other things being equal, female lifetime reproductive success (LRS) should be compromised where conditions deviate from monogamy. As developed in 1972 by Trivers, a male's LRS is a function of the number of successful implantations (also see Alexander *et al.*, 1997; Rice, 2000). In general, anisogamy will favor males who adopt tactics and strategies to minimize the likelihood of multiple mating by females and to maximize females' "proximate fecundity" since sperm competition is deleterious to a male's LRS (Rice, 2000; Alexander *et al.*, 1997). In contrast, females will be selected to optimize long-term over short-term fecundity and to mate-multiply to the degree that it benefits their LRS (Rice, 2000; Dixson, 1998; Jones and Cortés-Ortiz, 1998). As argued in Chapters 5 and 6 of this volume, these opposing interests and programs derive, in part, from the differences between males and females in their confidence of r, a state deriving fundamentally from anisogamy and the environmental potential for polygyny (see Wittenberger, 1980; Nunn, 2003; Horwich et al., 2001; Jones, 2004).

that antagonistic coevolution is most likely to occur in multimale–multifemale (polygynandrous) structures and/or where the likelihood of multiple mating by females is high, the same conditions thought to favor behavioral flexibility (see Chapters 1, 2, 8). Before reviewing the range of sociosexual organization found in primates and discussing these in relation to environmental heterogeneity, including the generation of behaviorally flexible responses to intergenomic sexual conflict, this chapter will attempt to show that several poorly understood features of primate behavior may be interpreted productively as antagonistically coevolved responses. Haldane (1949 cited in Summers *et al.*, 2003, p. 640) argued that mutually antagonistic interactions may "generate diversity both within and between species."

Does Each Sex Favor Different Outcomes
of Male–Female Interactions?

No empirical documentation exists to test the possibility that optimal outcomes differ for male and female primates. Using "agent-based" modeling and genetic algorithms applied to antagonistic coevolution, however, Nunn (2003) tested the question, "How can a female entice a male to mate, given that other females in the group may be cycling and mating effort is costly for males?" Running these simulations with and without the possibility of infanticide, Nunn's results supported the view that sexual conflict leads to antagonistic coevolution between males and females, an effect that was stronger with infanticide permitted.

These treatments highlight the important role that theoretical formulations can play for primatologists and support the view that flexible behaviors (e.g., infanticide) may confer an advantage to one sex at the expense of the other, a necessary condition for the demonstration of an evolutionary arms race (Chapman *et al.*, 2003a). Comparative studies, also discussed by Nunn (2003), are additionally useful for identifying congenerics for which the optima for various characters change over evolutionary time. Primate genera in which adults of both sex exhibit color at the same location on their bodies might be excellent candidates for such an analysis (Gerald, 2003). Of course, for primates, unlike many other taxa (e.g., Arnqvist and Rowe, 2002), the functions of signals are rarely known (Shahnoor and Jones, 2003; Gerald, 2003), another area of investigation in need of systematic study. Chapman *et al.* (2003a, p. 43) point out that "another powerful means of demonstrating the existence of sexual conflict and the function of the underlying traits involved is to manipulate them genetically" in order to measure the costs and benefits (e.g., differential mortality) of characters. Primatologists have historically been resistant to such experiments. However, it might be justified to conduct tests of antagonistic coevolution with genetic manipulations with populations and/or species that are polygynandrous, not endangered, and have a relatively brief lifespan (e.g., *Saimiri sciureus*; cooperatively breeding marmosets and tamarins). Alternatively, government agencies might be lobbied to fund longitudinal primate studies that would overlap the lifetimes of human researchers.

A General Formulation for Antagonistic Coevolution
between Males and Females

An apparent paradox in studies of antagonistic coevolution is "the phenomenon of male-induced harm to their mates" (Rice, 2000, p. 12953;

Johnstone and Keller, 2000; Watson *et al.*, 1998; Royle *et al.*, 2002). How might males benefit from depressing the reproductive success of females? This question is fundamental to an investigation of the costs of mating for mammalian males and females. Following the arguments of several authors (Schoener, 1971; Trivers, 1972; Shahnoor and Jones, 2003, p. 20), males are expected to apportion relatively more of their "fitness budget" to mating effort, all other things being equal, than are females because initial male investment (sperm) in reproduction is less in males than in females, because variance in male reproductive success will generally be greater for males than for females, and because males will generally have more to gain *and* to lose from mating than females in the same conditions. Females should be willing to apportion a greater relative amount of their "fitness budget" to parenting effort because of higher relative investment in young (eggs) compared to males and because variance in reproductive success for females will be lower than that for males, *ceteris paribus*. These conditions obtain particularly for mammalian species, including primates, whose females generally assume costs of gestation and parenting without noteworthy investments of time and energy by males. As suggested above, however, arguments based upon anisogamy subject to confirmation where sexual conflict may be operating.

Discussing differences between males and females in "optimal fecundity and remating rates," Rice (2000, p. 12953) advances a schema whereby a mutation occurs at one locus increasing male fitness but decreasing female fitness. This "male-gain/mate-harm allele" will "fix" in the population independent of its effects upon females if the mutation is sex-limited (male) or if it is not counterbalanced by expression in the female sex. The next stage in a cycle of antagonistic coevolution would be a comparable response (intergenomic conflict) by females at a separate locus, reducing the costs of the "male-gain/mate-harm allele." Where interactions between the sexes are biased by anisogamy, most cycles of antagonistic coevolution will be initiated by males. Rice (2000, p. 12953) posits, further, that where the correlation between male and female traits is <1, "males can evolve traits that harm their mates."

An inference from these formulations is that, all other things being equal, it is always costlier for a female to mate with a male than for a male to mate with a female, in part because it will benefit males to parasitize females in an attempt to obtain additional reproductive benefits (Alexander *et al.*, 1997; Gowaty, 1997). This likelihood further reinforces the argument in Chapter 5 that it will often be beneficial for females to adopt a conservative suite of behavioral tactics and strategies compared to males as a counterstrategy to potentially harmful effects by males (Rice, 2000). As Figure 7.1 suggests, behavioral flexibility may be particularly costly to females in heterogeneous regimes if they expose her to increased exploitation by males (sexual conflict). It seems logical to suggest, however, that behaviorally flexible tactics and strategies

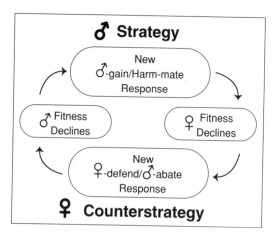

Fig. 7.2. Generalized Rice model showing intersexual antagonistic coevolution between interacting phenotypes [based with permission upon ©Rice, (2000, Figure 2, p. 12954)].

by females will be favored wherever the genetic and/or phenotypic benefits outweigh the costs of "male-gain/mate-harm" responses and/or where they decrease these costs below some threshold value. A corollary of this idea is that males are expected to demonstrate tolerance, patience, persistence, and/or "aggressive restraint" to a prospective mate if the (genetic) benefits of doing so outweigh the costs (Jones, 1996b; East *et al.*, 2003; Moore *et al.*, 2001; Forsgren, 1997; Royle *et al.*, 2002).

Figure 7.2 generalizes Rice's (2000) pictorial representation of an antagonistic coevolutionary cycle in order that the exhibit include mechanisms of behavioral flexibility other than mutation (e.g., learned responses with underlying genetic variation; see Chapter 1, Table 1.1). Further assumptions of this graphic are identical to those of Rice (2000) described above. Although there is no empirical evidence for primates demonstrating the existence of "male-gain/mate-harm alleles" or that the tactics and strategies resulting from these responses are beneficial to male survival and/or reproductive success, copious data exist showing that primate males may damage females reproductively, both directly (via force) and indirectly (via infanticide; Smuts and Smuts, 1993; van Schaik and Janson, 2000; Crockett, 2003; Palombit, 2003; Jones and Agoramoorthy, 2003), a condition for the initiation of an antagonistic coevolutionary cycle between the sexes such as that displayed in Figure 7.2. In the following sections, I will consider four behavioral responses—one by males ("rape"), and three by females (multiple-mating, "female dominance," and "homosexuality" in females)—in relation to intersexual antagonistic coevolution and the expression of behavioral flexibility.

The Extent and Limits of Extreme Selfishness: Forced Copulations by Males as an Indicator of Sexual Conflict

Coercive mating (Smuts and Smuts, 1993), a form of intersexual selection like all sexually selected events (West-Eberhard, 1979), implies the acquisition of mates by "intimidation, harassment, and/or physical force" (Thornhill and Palmer, 2000; Smuts and Smuts, 1993). Coercive mating, however, might entail any mechanism of mate acquisition in the (genetic) interests of one sex, costly to the fitness of the opposite sex (e.g., social parasitism; Jones, 1997a). It is in this broader sense that coercive mating can be viewed in the context of sexual conflict and antagonistic coevolution, a view presaged by Smuts and Smuts (1993; also see Thornhill and Palmer, 2000; Nunn, 2003). With the exception of male humans (*H. sapiens*; Thornhill and Palmer, 2000) and subadult male orangutans (*Pongo pyugmaeus*; Rodman and Mitani, 1987), reports of "rape" (forced copulation) in primates are infrequent (Smuts and Smuts, 1993; Dixson, 1998; Smuts *et al.*, 1987; Jones, 2002b), and Dixson (1998) suggests that, in general, primate rape may be pathological. The broader interpretations of unsolicited matings advanced by Smuts and Smuts (1993), Thornhill and Palmer (2000), and the present treatment, however, widen the scope of this topic to include numerous responses of prosimians, monkeys, and apes, including humans.

In mantled howler monkeys (*A. palliata*) force is rarely employed by males to achieve copulation. A male forcefully or aggressively approaching a female with the intent (motivation) to mount is almost always successfully repulsed by the female's open-mouth bared-teeth display, generally accompanied by vocalizations (Jones, 1985a). The existence of these condition-dependent behavioral responses might be interpreted in accord with a traditional ethological explanation as stereotyped or ritualized intention movements, ambivalent responses, or redirected acts (Tinbergen, 1952) to resolve conflicts of interest and avoid the risks of serious injury that may result from costly fights (Krebs and Davies, 1993). However, based upon the results of computer simulations of monogamous mating systems, Wachtmeister and Enquist (2000) have recently proposed that stereotyped or ritualized "displays" derive from intersexual conflict. If these results can be shown for polygynous, polygynandrous, and/or polyandrous taxa, they may contribute to an understanding of the evolution of stereotyped and ritualized signals and displays in primates (see, for example, Smuts and Watanabe, 1990; Jones 2000; Chapter 6). Wachtmeister and Enquist's (2000) studies depend upon the assumption that males exploit the response biases of females (also see Chapman *et al.*, 2003b), an area of research that is virtually unexplored in primatology.

Wachtmeister and Enquist's (2000) formulation is consistent with Figure 7.2 since, in their simulation, antagonistic coevolution is modeled as an initial male "display" represented by a sequence of signals "exploiting" a female recognition mechanism represented by an artificial neural network. In polygynandrous mantled howler monkeys (*Alouatta palliata*), males are thought to employ stereotyped and ritualized "compound displays" as condition-dependent signals of attraction and quality (Jones, 1999c, 2002c, 2003a; Chapter 6), possibly as a counterstrategy to the ritualized open-mouth bared-teeth display of females, which might best be interpreted as a counterstrategy to an original condition in which some proportion of males in a population was likely to exhibit forced copulation (Jones 2002b). This interpretation is consistent with Wachtmeister and Enquist's (2000) artificial demonstration that females "evolve" counterstrategies to male behaviors entailing increased resistance to male exploitation (also see Jones, 1997a, 2002a). Condition-dependent displays by *A. palliata* males, then, may exploit females' sensory (including reward) systems, decreasing their resistance to copulation, a hypothesis consonant with that of Wachtmeister and Enquist (2000; also see Jones, 2002c), and one deserving empirical investigation in this and other species of social vertebrates.

Once selection has led to decreased resistance to copulation in females, the stage has been set to favor multiple mating by females and subsequent sperm competition (Harcourt, 1998; Jones, 2002b), events deleterious to male fitness which may be ubiquitous in the primate order (Jones and Agoramoorthy, 2003; Dixson, 1998). No evidence exists for primates to test the possibility that any of the behaviors highlighted in this discussion (e.g., stereotyped and ritualized signals and displays, forced copulation, multiple mating) are most common in the repertoires of species found in heterogeneous regimes, associations that would be predicted by the schema outlined in Figure 7.1. Citing theoretical and empirical studies, however, Koenig (2002) points out that indicators of behavioral conflict are more likely in unpredictable regimes.

Multiple Mating by Females as a Counterstrategy to Male Infanticide

The intensity of sexual selection varies in time and space with environmental heterogeneity (Emlen and Oring, 1977; Andersson, 1994; Shuster and Wade, 2003), and multiple mating by females is generally associated with levels of abiotic and/or biotic change. Wolff and Macdonald (2004) persuasively argued that Hrdy's (1979) hypothesis for the evolution of multimale mating (MMM) by mammalian females is correct. Hrdy suggested that "multimale mating functions to confuse paternity, which, in turn, deters infanticide" (Wolff and Macdonald, 2004, p. 127). Evaluating data for the 133 species of mammals in which MMM has been documented, Wolff and Macdonand's (2004, p. 128, Table 1) analysis strongly indicates that, of the nine hypotheses attempting to explain MMM, Hrdy's explains most of the observed variance in the data set. Further, in box 2 of their paper (p. 129), Wolff and Macdonald (2004, p. 131) describe infanticide as a "pacemaker" for the evolution of MMM based upon the "scenario" of van Schaik and Kappeler (1997), showing "a possible transition from polygyny to promiscuity and monogamy." Although Wolff and Macdonald (2004) place little emphasis upon thresholds of response and do not discuss operational sex ratios as both of these might relate to the causes and consequences of infanticide (see Queller, 1997), this review and its conclusions provide a very tight case for the utility of Hrdy's treatment given available research, both theoretical and empirical.

Wolff and Macdonald (2004) argue that the *origin* of MMM was not sexually selected (i.e., it did not occur in response to genetic or other benefits derived from mating with one male over another). Instead, these authors consider the benefits of MMM to be derived wholly from those gained by a female in protecting her living offspring from infanticide. Wolff and Macdonald (2004, p. 130) do point out that sexually selected benefits might be obtained secondarily "once MMM evolved for some other purpose." In light of other treatments of "promiscuity," it will be necessary to carefully unpack the relationship, if any, between MMM and sexual selection. Holland and Rice (1999), for example, demonstrate a relationship between sexual selection, "promiscuity," and "intersexual antagonistic coevolution." Hrdy (1974, 1979; also see van Schaik and Janson, 2000) originally claimed that infanticide by males was sexually selected and, consistent with Holland and Rice's (1999) arguments; MMM might be viewed as a response to intersexual conflict and a male trait (infanticide) increasing male fitness at a female's expense. To quote Holland and Rice (1999, p. 5083), "Conflict between mates hinges on sexual infidelity.... [W]henever an individual has multiple mates, the lifetime reproductive success of that individual will differ from the success of its mates. Thus, promiscuity necessarily introduces the opportunity for sexual conflict through the evolution of novel traits that increase the reproductive success of members of one sex at a cost to members of the opposite sex." Multimale mating, probably a ubiquitous trait among primates, implies intersexual conflict, a topic in the early stages of investigation for the Primate Order.

Research on MMM by primate females and those of other taxa has the potential to contribute to the formulation of general principles of social behavior. Wolff and Macdonald's (2004) discussion, for example, may be viewed as a description of one class of transactions important to potential mates among social mammals and other groups of organisms. Shellman-Reeve and Reeve (2000) (mathematically) model interactions between males and females in accord with transaction theory, a category of reproductive skew models including "concession" and "constraint" models (see Hager, 2003). In this view, promiscuity is a "transaction between social mates" (Shellman-Reeve and Reeve, 2000, p. 2543).

Females and males, then, are assumed to be in conflict over the most beneficial tradeoff of mating effort and parenting effort. Neff (2001) points out that, where males are larger than females (as for most primates), an amended "tug-of-war" model (Clutton-Brock, 1998; Hager, 2003) may be most appropriate, and he describes Shellman-Reeve and Reeve's (2000) treatment as showing that "the evolution of infidelity" (Neff, 2001, p. 175) requires an assessment of both female and male interests. Because Shellman-Reeve and Reeve (2000) model these states elegantly and with general import, Neff (2001, p. 175) suggests that "transactional theory might provide the basis for a truly unifying theory of social interactions."

Several recent publications (e.g., Neff, 2001; Reeve, 2001) have argued that theories of "reproductive skew" may yield general formulations for the evolution of social behavior. There is some disagreement, however, about the relative utility of "transactional" models, on the one hand, and "tug-of-war" models ("indirect control" models), on the other. In a recent paper, Langer *et al.* (2004) have tested the predictions of these two categories of models with the social bee, *Exoneura nigrescens.* Consistent with Clutton-Brock's (1998) arguments, "tug-of-war" models are supported. This finding is in accord with that of Widdig *et al.* (2004, p. 819) who provided evidence for the view that "reproductive skew in male rhesus macaques is best accounted for by the "limited control" model, with multiple factors interacting to regulate individual reproductive output." Additional research is required to assess models of reproductive skew for female primates and for species in addition to *M. mulatta.* It will also be critical to investigate the relationship between reproductive skew and environmental heterogeneity since the issue of "ecological constraints" is fundamental to these models (e.g., Reeve and Emlen, 2000).

"Female Dominance" in Primates: Counterstrategies that Benefit Females

Intersexual interactions may significantly impact social relations within primate groups and may lead to sociosexual organizations in which females are consistently more aggressive than males, winning most contests for limiting resources (e.g., gray mouse lemurs; Radespiel and Zimmermann, 2001; Radespiel *et al.*, 1998). Radespiel and Zimmermann (2001) have suggested that female dominance in gray lemurs and other lemurids is associated with environmental heterogeneity, in particular, the severe conditions associated with the climate of Madagascar. Assuming that female dominance is a derived characteristic among these prosimians and other primate species in which the trait has been identified (e.g., Kappeler, 1993; Boinski, 1999; Vervaecke *et al.*, 2003), heterogeneous local conditions may depress female reproductive success below some threshold value, favoring males who defer to females and exhibit "aggressive restraint" (Jones, 1982b, 1996b,c, 1997a, 2002c).

In extremely heterogeneous conditions, then, all other things being equal, "female-gain/mate-harm" alleles may be favored over "male-gain/mate-harm" alleles. This possibility further supports Pagel's (2003) conclusion that relations between the sexes may not always be a straightforward function of anisogamy—an initial condition from which female dominance would not be predicted. In Figure 7.1, female dominance might be expected at the extreme right of the continuum, beyond which point males might actually be resistant to or unwilling to mate (see, for example, Vasey, 1998, 2002). How might "same sex partner preference" or flexible "homosexual" behavior optimize benefits to an individual's genotype and/or phenotype?

"Same Sex Partner Preference" and Antagonistic Coevolution

Selection may act on males and females differently because traits increasing the fitness of one sex may decrease the fitness of the opposite sex (see Trivers, 1972; Rice, 2000). Studies of homosexual behavior in primates indicate that it is a behaviorally flexible, condition-dependent response (Vasey, 1995; Roughgarden, 2004). Observers of certain primate populations (see Chasteen, 2003; Morton, 2003) have recorded behaviors claimed to have no adaptive significance (e.g., "same sex partner preference"; Vasey, 1998, 2002) or to be maladaptive ("homosexuality", Dixson, 1998; infanticide, Dixson, 1998). Considering contrasts between predictions of sexual selection theory and sexual conflict theory, Gavrilets *et al.* (2001, p. 272) concluded that "female mate choice generated by sexual conflict is ... expected to lower overall population fitness and ultimately increase the risk of extinction." This condition may arise because sexual conflict may drive males, on the one hand, and females, on the other, to exhibit extreme resistance to mating. The theoretical treatment by Gavrilets and his colleagues shows under what circumstances these conditions are likely to arise. Most notably for the present discussion, female behavior is expected to be driven by the costs of mating (e.g., harassment by males and, perhaps, other females, mating rates, the timing and duration of mating, effects of male ejaculates, conflicts over differential parenting effort, in addition to the costs of resistance; see Andersson, 1994). According to Gavrilets *et al.* (2001), female resistance is most beneficial where males are selected to coerce females rather than to stimulate them.

Vasey (quoted in Chasteen, 2003) proposed that same sex partner preference by some female Japanese macaques (*M. fuscata*; see Chapter 2) has evolved in response to reticence to mate by males of this species, and Soltis and his colleagues (1997a,b) have found that female choice may influence patterns of mating in this species more than male coercion. Female Japanese macaques, then, while not demonstrating female dominance, may be preadapted to exert an effective degree of control over males and, perhaps, to exhibit relative "emancipation" from male control in their behavioral decisions. Japanese macaques are found in heterogeneous regimes (F. Fukuda, personal communication), and the observations reported by Vasey and Soltis *et al.* are consistent with sexual conflict theory and the idea that, above some threshold value of spatio-temporal unpredictability of females, males gain more by exhibiting aggressive or coercive restraint toward females, a condition that would favor tactics and strategies of female control, emancipation, or exploitation of males (Jones, 1996c, 1997a; Jones and Cortés-Ortiz, 1997).

As suggested above, beyond some threshold value the costs of male coercion and, in the extreme, mating itself, may outweigh their benefits to males, perpetuating cycles of antagonistic coevolution. Costs to males may represent, for example, costs in time to attract resistant and/or "choosy" females, female condition, timing or context of mating, or costs incurred from multiple mating by females, constraints imposed by exploitation of females' sensory systems, behaviors that may be most common in multimale–multifemale societies such as those of Japanese macaques. In these conditions, thresholds of response to female stimuli (e.g., olfactory, visual, tacticle) may decrease, with the consequence that some males may not find females attractive. The initial effect of these responses would be to restrict gene flow, as may be occurring in some of the groups studied by Vasey (1998, 2002). It would be important to identify optimal male and female mating patterns in these populations compared to populations not demonstrating same sex partner preference by females in order to test the hypothesis that reproductive barriers are occurring, possibly in response to sexual conflict. Related to this, it will be necessary to investigate the particular regimes in which Vasey's subjects are found (troops outside Kyoto isolated from other troops of the same species and provisioned for many years, in populations from which males rarely emigrate or into which males rarely immigrate; F. Fukuda, personal communication).

Patterns of "homosexual" behavior across the Order Primates support the above scenario. Vasey (1995; also see Bagemihl, 1999) found, for example, that homosexual behavior is absent among lemurids. This finding suggests that female dominance may have evolved as a counterstrategy by females to coerce males to mate. Differences might obtain, also, in relative strengths of sexual conflict and natural selection (see Gavrilets *et al.*, 2001), in relative differences between male and female optima, or the impact of other factors on the fitness of males and females.

In an important study, Ginther and her colleagues (2001) demonstrated that captive mature male cottontop tamarins (*Saguinus oedipus oedipus*) exhibited "restricted" sexual behavior with females of their natal group but relatively high levels of sexual behavior with other males of their group. Mounting behavior, including visual and vocal signals, between males was indistinguishable ethologically from mounting behavior between adult males and females. These authors suggest that "sexual behaviour directed at other males" (i.e., homosexual behavior) may represent "functional suppression" of (heterosexual) reproductive behavior, a proximate explanation. Although homosexual behavior may be a straightforward case of inbreeding avoidance, the immediate consequence of this phenomenon is expected to promote the restriction of gene flow while the mid- or long-term effect may be to induce emigration by group members unable to find opposite sex mates in the group, an effect that would promote gene flow. Similarly, individuals of either sex may immigrate into the group, also promoting gene flow. These immigrating individuals may exhibit genotypic and/or phenotypic traits different from group members that would increase likelihoods of heterosexual matings. It is expected that same sex partner preference will be most likely to occur in heterogeneous regimes, conditions that may promote female dispersal in species with typically philopatric females or that may increase rates of dispersal by one or both sex (see Chapter 2). In order to test these possibilities, field studies, including field experiments, are required.

Same-Sex Partner Preference by Females as a Strategy for Managing Male Power

Roughgarden has suggested (quoted in Chasteen, 2003; also see Gewin, 2003) that, "Female choice...has much more to do with managing male power than it does with trying to obtain good genes," implying that altered mating decisions by females may negatively impact and constrain the reproductive "choices" of males. She believes that a theory of sexuality should incorporate social as well as sexual selection. As pointed out by Crook (1972; also see West-Eberhard, 1979, 1983; Jones and Agoramoorthy, 2003; Crook and Gartlan, 1966), social selection may explain many of the traits characteristic of males in multimale–multifemale societies varying within (and between) individuals in time and space and corresponding to differential reproductive success [also see "competition-dependent" processes discussed by Gross (1996), West-Eberhard (1979), and Jones and Agoramoorthy (2003)]. West-Eberhard (1979, p. 223) defined social selection as "selection involving direct competition via social interaction," pointing out that Darwin (1871) was concerned broadly with social competition, one category of which was competition for mates. In this broader sense, then, behavioral flexibility is thought to occur in response to density-dependent and/or frequency-dependent interactions (e.g., in response to interaction rates) between individuals (and groups?; see Koenig, 2002) in spatial and temporal regimes that may change over time (Jones and Agoramoorthy, 2003), and, in part for this reason, demographic factors are incorporated into the models of Gross (1996), discussing alternative reproductive behaviors, and Gavrilets (2000), discussing reproductive barriers and sexual conflict (also see West-Eberhard, 1983).

Insights into these topics are gained by a consideration of Vasey's (2000) experimental manipulation of sex ratios in *M. fuscata*. In the laboratory,

Vasey (2000) manipulated the proportion of males and females in groups of Japanese macaques in order to assess the comparative frequency of "homosexual activity" in baseline (sex ratio typical of species) and experimental (sex ratio skewed toward females) groups. Similar to previous findings from naturalistic studies cited in his paper, Vasey (2000, p. 17) found that, "compared to the baseline period, females solicited significantly more same-sex individuals for sex and formed significantly more homosexual consortships during the experimental period of the study." Vasey concluded that same sex partner preference in experimental groups was not a function of the absence of "heterosexual alternatives" and that increased levels of homosexual behavior were a function of increased sexual opportunities with females occasioned by the increased numbers of same-sex individuals in the experimental group. The latter inferences have been made consistently in the theoretical and empirical literature on sexual selection whereby alternative sexual phenotypes have been shown to be a function of variations in operational sex ratio (Andersson, 1994; Shuster and Wade, 2003), a conclusion drawn by Andres *et al.* (2003) in their study of alternative reproductive strategies in gray mouse lemurs (*Microcebus murinus*).

It might have been advisable for Vasey (2000) to view his homosexual or bisexual female subjects as male mimics and/or social parasites of males, competing with them for access to receptive females, thereby biasing the *effective* operational sex ratio (OSR) in his experimental group toward (ostensible) males rather than toward females. The consequence of this recalculation would be to increase rather than decrease competition for receptive females. Female homosexuality or bisexuality (female mimicry of male behavior, see Table 1.1) may be, then, a counterstrategy to conditions in which male–male competition for females is decreased (OSR's favoring the female sex), increasing *effective* male control over females by increasing the representation of the male phenotype within the breeding unit. In this view, same sex partner preference by females may represent a reproductive tactic or strategy to manipulate competitive relations among biological males whose effects would be to increase "power" and/or control by Japanese macaque females (see Beekman *et al.*, 2003; Jones, 2000, 1997a). It is in this sense that female selectivity is related to "managing male power" in Japanese macaques and other primate species in which females retain "the ability to do or act in a situation in which conflict over reproduction exists" (Beekman *et al.*, 2003, p. 277).

Conclusions

If "the ability to control reproduction when conflict exists" (Beekman *et al.*, 2003, p. 277) is more problematic for individuals in heterogeneous regimes such as those inhabited by Japanese macaques (Fukuda, 2004) or mantled howlers (Jones, 2000), behaviorally flexible responses may be employed to increase the likelihood of (reproductive) control by one individual (or group) over others. In a sense, then, all reproductive conflicts between individuals are conflicts over relative influence [for ultimate genetic gain (e.g., "reproductive skew"; Hager, 2003a,b)], and it is logical to assume that individuals will benefit by having responses in their behavioral toolbox that increase their power relative to others. This chapter has argued that intersexual conflict may favor behavioral flexibility by individuals, particularly in temporally and spatially heterogeneous regimes. Chapter 8 considers the particular types of sociosexual organization in the Order Primates, assessing these structures in relation to environmental predictability, the dispersion and quality of limiting resources, and the implications of these factors for the decisions that individuals make for the optimization of power relations, inclusive fitness, and phenotypic success.

Sociosexual Organization 8
and the Expression of
Behavioral Flexibility

[T]he same conceptual framework can be used to study the social organization of insect and vertebrate societies. Ecological factors, together with internal factors, such as relatedness, determine the degree of within-group conflict, partitioning of reproduction and the stable social structure of animals, independently of whether they are ants, birds, or mammals.

Keller (1995, p. 359)

Introduction

Most vertebrate populations are structured, and population structure will be an emergent property of decisions made by individuals concerning where to reside and where to reproduce. Individuals of mammalian species, then, are generally not organized randomly with respect to features of the habitat or to one another (but see Caughley, 1964). Students of social organization seek to explain patterns of interindividual organization within the framework of organismic and evolutionary biology and to identify and measure the causes and effects of population dispersions. For most species of mammals, including primates, the determinants of population distribution and abundance are poorly understood. However, most investigators assume that these patterns are a function of the dispersion and quality of limiting resources (e.g., food, mates), dispersal costs (Johnson *et al.*, 2003), as well as pressures from predation (Dunbar, 1988; Sterck *et al.*, 1997; Nunn, 2003; also see Smuts *et al.*, 1987).

Population structure may constrain the potential for behavioral flexibility through temporal and spatial heterogeneity in the frequency, rate, duration,

109

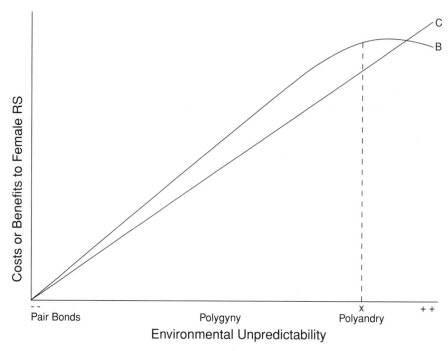

Fig. 8.1. A graphical model describing the costs (C) or benefits (B) to female reproductive success (RS) of sociosexual assemblages as a function of environmental unpredictability. Benefits will increase and then level off as the costs increase linearly (because
resources are limiting), and the maximum net benefit (benefit minus cost) to females
should occur at "x." The location of "x" will depend upon the position and shape of the
benefit and cost curves, a function of environmental unpredictability over the short and
long terms (after Jones and Cortés-Ortiz, 1998).

and intensity of interindividual interactions. All other things being equal, it
is in the large multimale–multifemale primate assemblages that behavioral
flexibility (e.g., in the treatment and processing of food and in sociosexual
relations) is most noteworthy (e.g., *Saimiri sciureus, Cebus capucinus, Alouatta
palliata, Macaca fuscata, Papio cynocephalus, Pan* spp., *Papio* spp., *Homo sapiens*).
Sexual selection is expected to be intense in these taxa (e.g., Harcourt, 1998),
and theoretical work has shown that the intensity of sexual selection will be
positively associated with niche breadth (Proulx, 1999), a condition expected
to favor behavioral flexibility. Following schemas proposed in Chapter 1,
Figure 8.1 displays the expected optimal sociosexual structure for females
as a function of environmental heterogeneity. There may be more than one
route to behavioral flexibility, however, since environmental heterogeneity
may derive from more than one source (e.g., heterogeneity within habitats
or heterogeneity between habitats). Cooperatively breeding callitrichids, for

example, display a noteworthy degree of behavioral flexibility although they typically live in small groups compared to multimale–multifemale primates (see Saltzman, 2003).

Environmental and Phylogenetic Constraints on Behavioral Flexibility

As a preliminary test of the proposed causal links between environmental heterogeneity and behavioral flexibility, the characteristics of three allopatric species of howler monkeys living in three different environments were assessed. Table 8.1 compares several morphological, physiological, and behavioral traits for A. palliata (the mantled howler monkey), A. seniculus (the red howler monkey), and A. caraya (the black and gold howler monkey). The species are ordered according to features of the habitats which they occupy, especially, their relative tolerances for secondary forest growth and xeric conditions, from least tolerant to most tolerant. Among the characters stable at the generic level, and, therefore, nonheritable, the age-reversed dominance system (Jones, 1978, 1980, 1983a) is especially noteworthy since it suggests relatively stable developmental canalization of interference responses, including signals and displays. Table 8.1 also shows that other traits may be stable or variable at the generic level, for example, feeding strategy, dispersal patterns, and adult size dimorphism. Differences between species, however, appear to be influenced primarily by differential habitat factors in directions consistent with the schema presented in Chapter 1. For example, single male and "age-graded" groups are most characteristic of A. seniculus and A. caraya, those species inhabiting the more xeric conditions in which patch variation (distribution, abundance, and quality) is low, while the most typical sociosexual dispersion for A. palliata is the multimale–multifemale structure most often associated with habitats in which patch variation is high (Emlen and Oring, 1977; Roughgarden, 1979; Wittenberger, 1980).

While mantled howlers and, to a greater degree, black and gold howlers are relatively specialized for habitat type compared to red howlers, the food quality and dispersion in forests preferred by mantled howlers favor a larger range in group size, larger ratios of adult males to adult females in a group, the signaling of estrus condition to group males (either to "incite" male–male competition or to attract mates; Jones, 1985a, 1997a, 2003a), relatively low copulation frequency, implying selectivity by males as well as females (Jones, 1985a), greater size dimorphism between adult males and adult females, and exaggerated testis size and color relative to body size and color (consistent with a greater intensity of sperm competition likely in multimale–multifemale groups; Jones, 1999c; Jones and Cortés-Ortiz, 1998; see Reeder, 2003). All of these differences may favor the expression of flexible behaviors by exposing

Table 8.1. Comparative Summary of Characteristics for Three *Alouatta* Species. It is Assumed that Variations in Heritable Responses may Lead to Behavioral Flexibility within, as well as between, Individuals.

Character	*Alouatta*[a]		
	Palliata	*Seniculus*	*Caraya*
Population density/km^2	67	108	25
Tolerance for secondary growth under xeric conditions	Low	Moderate	High
Food	Herbivore/folivore (hf)	Hf	Hf
Drinks water	Rare	Rare to occasional	Rare
Moves on ground	Occasional	Yes	Occasional
Sociosexual organization	Polygynous, age-graded, multimale–multifemale (modal multimale–multifemale)	Same (modal polygynous)	Same (modal polygynous)
Adult color dimorphism	No	No	Yes
Adult size dimorphism	77%	79%	79%
Adult sex ratio (males:females)	2.4	1.6	1.0
Infant "natal coat"	Yes	Slight	Yes
Dispersal x sex	Males and females	Males and females	Males and females
Vulval tumescence with estrus	Yes	No	No (captive)
Vulval color change with estrus	Yes	Slight or none	Slight
Menstrual blood visible	No	No	Slight (field and captive)
Copulation position	Variable dorso-ventral	Same	Invariant dorso-ventral (field and captive)
Copulation frequency	Low, about 0.08/h	"Moderate"	Relatively high, about 0.30/h (captive)
Lingual gesture	Yes, about 2/sec	yes ("uncommon")	Yes, about 1/sec ("uncommon")
Lingual gesture limited to sexual context	Yes	No, sometimes in aggressive context	No
Consort behavior	Yes, uncommon	Yes	Yes

Table 8.1. (*continued*)

Harassment and interruption of mating pairs	Yes, especially by juveniles and adult females	Not direct, but indirect by displacement	Yes (captive)
Dominance system	Age-reversed (AR), adult males and females	AR, adult females	AR, adult females
Coalition displacements	Yes	Yes	Yes
Scrotal and vulval displays	Yes	?	Yes
Penile display	Yes	Yes	Yes
Clitoral display	No	Yes	Yes
Branch-break display	Yes	Yes	Yes
Piloerection (adults)	yes (females?)	Yes	No
Infant transfer	Yes	Yes	Yes
Agonistic buffering	Yes, rare		Yes (captive)
Infanticide by adult males	Yes	Yes	Yes (field and captive)
Olfactory communication	Apparently common	Apparently common	Apparently uncommon (captive)
Perianal marking	Yes	Yes	Yes
Chin-rubbing	Yes	Yes	Yes
Urine wash	Yes	No	No
Grooming (tactile communication)	Rare (about 0.09 events/h, deciduous habitat; about 0.03/h, riparian habitat)	Common (about 0.36/h; llanos)	Common (about 6.4/h; captive)
Direction of grooming	Dominants groom subordinates	Variable	Variable
Grooming solicitation	No	Yes	Yes
Greeting ceremony	Yes		Yes (captive)
Huddling	Yes	Yes	Yes
Bridging	Yes	Yes	Yes (captive)
Vocal communication	Common	Common	Common (captive)
Percent (%) leaves in diet	54	53	79

[a] Information in table extracted from Carpenter 1934, 1965; Colillas and Coppo, 1978; Crockett, 2003; Crockett and Eisenberg, 1987; Eisenberg *et al.*, 1972; Glander, 1975, 1980; Hrdy, 1979; Jones, 1978, 1979, 1980, 1983a, 1985a, 1996c, 1997d, 2003a; Milton, 1980; Neville, 1972a,b; Rudran, 1979; Thorington *et al.*, 1979. Personal communication from C. Bell, C. Crockett, M. Neville, R. Rudran, A. Shoemaker, and G. Zunino is gratefully acknowledged.

an individual's phenotype to a greater range of environmental conditions and potential stressors.

Other characteristics of the species in Table 8.1 also appear to follow from physical and biotic parameters, factors likely to influence the decisions made by individuals in these regimes. For example, *A. seniculus*, the least habitat specialized of these three species, occupies the largest geographical range and exhibits the highest population density. Similarly, the relative benefits and costs of differential specialization are apparent when *A. palliata* and *A. caraya* are compared. While the black and gold howler monkey exhibits the lowest population density of these three species and displays significantly greater specialization for folivory compared to mantled howlers, *A. caraya* is successful over a greater biogeographical range than *A. palliata*, particularly in xeric conditions, which mantled howlers cannot tolerate and red howlers do not favor. A greater dependence upon leaves may account for the wider geographical distribution of black and gold compared to mantled howlers. Although behavioral flexibility may be associated with breadth of geographical range where the tempo and mode of environmental heterogeneity relative to T (generation time) occurs below some threshold value, the black and gold howler monkey is a monotypic species, a characteristic often associated with extremely specialized types. *A. caraya* is a relatively greater food and habitat specialist than *A. palliata* whose diet is more catholic, implying that black and gold howlers utilize a narrower "resource gradient" than mantled howlers. For the traits displayed in Table 8.1, *A. seniculus* exhibits the greatest generalization, an observation consistent with this species' wide geographical distribution (Jones, 1997c) and indicating, perhaps, that it is ancestral to both *A. palliata* and *A. caraya*, the (relatively) more specialized types.

Miller (1956) suggested that polytypic species (e.g., *A. palliata*) would be neither too restricted in their habits nor too eurytopic (capable of tolerating a wide range of environmental changes), since adaptation to a narrow spectrum of conditions would necessarily limit a species' geographic or ecologic range within those tolerable limits and would retard polytypic differentiation by the very process of highly selective habitat preferences. Similarly, but at the other extreme, Miller (1956, p. 275) argued that "extremely eurytopic species with great tolerance and adaptability of individuals will not trend toward extreme polytypic variation." In these species, individual phenotypes would be so "plastic" that their behavioral compensations across a broad array of habitats would be flexible enough to neutralize potential differentiation, either by random or environmentally correlated effects (see Lewontin, 1957, 1974). According to Miller (1956, p. 274), the best colonizing species would be those that are "aggressive or flexible in the sense of range expansion, geographically or ecologically, and yet not too adaptable individually" since colonization attempts by individuals of these species "would set up new habits in the individual and throw the species under new selective influences so that racial evolution would be relatively rapid." It is expected, then, that the most highly differentiated species (e.g., A. *palliata*) will be neither the *most* (e.g., A. *seniculus*) nor the *least* (e.g., A. *caraya*) homeostatic forms.

Table 8.1 shows that A. *palliata*, the most polytypic species, is a herbivore generalist with a preference for a broad range of forest types (e.g., tropical wet, tropical moist, tropical dry) but with relatively restricted tolerance for very humid and xeric conditions. A. *seniculus*, on the other hand, is less differentiated compared to A. *palliata* and, consistent with Miller's analysis, more generalized in its geographical and ecological requirements (although relatively intolerant of xeric conditions). The most specialized species in our sample, A. *caraya*, is monotypic, as Miller's system predicts. Miller's model, consistent with the formulation of West-Eberhard (2003, p. 382), suggests, that behavioral flexibility (either generalist or polyspecialist; West-Eberhard 2003, p. 382) is most likely to be associated with an *intermediate* range of environmental gradients relative to T (e.g., A. *palliata*), that very high ranges of environmental gradients (relative to T) lead to a generalist strategy (A. *seniculus*), and that a narrow range of environmental gradients (relative to T) leads to a specialist strategy (A. *caraya*). Early reviews of *Alouatta* (Crockett and Eisenberg, 1987) stress the similarity of traits across species. The present treatment, however, suggests that detailed comparisons of differences across species should be investigated in order to test predictions concerning the relationship between environmental heterogeneity and behavioral flexibility. It is of interest that Perry (2003) reports that models of social learning predict the greatest proliferation of this mechanism in regimes with intermediate degrees of heterogeneity.

Sociosexual Organization in Primates: An Attempt at Classification

For Trivers (1985), "social" behavior may be "selfish," "cooperative," "altruistic," or "spiteful," two of these forms of interindividual interaction (selfish and spite) yielding negative outcomes on fitness for the recipient of the actor's behavior. Most schemas for the classification of societies, however, require that an actor's behavior assist the reproduction of a conspecific, and Wilson (1971, p. 469, 1975; also see Costa and Fitzgerald, 1996) has defined a "society" as "[a] group of individuals belonging to the same species and organized in a cooperative manner. Some amount of reciprocal communication among the members is implied." No primate study has attempted to assess the relative extent of positive (cooperation, altruism) or negative (selfishness, spite) interactions (relative to local conditions). However, Chapter 3 of this volume suggests that, where the environment permits, negative interactions, in which an actor's behaviors decrease the fitness of a conspecific recipient (particularly by manipulation and/or exploitation), may be very common in most primate species. I take this view because, similar to Zahavi's

(1975; also see Zahavi and Zahavi, 1997) treatments of "honest signaling," it seems reasonable to assume that individuals will behave selfishly whenever possible. Unlike Zahavi, however, it seems to me that an individual's selfish ends will often be thwarted (e.g., by social parasitism), leading, necessarily, to solitary, cooperative, or altruistic tactics and strategies. The extent to which primate societies retain individuals who are selfish and individualistic (see, for example, Nunn and Pereira, 2000; Jones and Agoramoorthy, 2003) is a measure of their failure to meet the criteria for the highest grades of sociality in the schemas devised by and derived from Michener and Wilson (the "M–W classification"; Costa and Fitzgerald 1996, p. 286; Crespi and Yanega, 1995).

Several authors have attempted to classify societies (see Costa and Fitzgerald, 1996; Crespi and Choe, 1997a,b). Most of these formulations have relied upon the influential M–W classification based upon features of the social insects, in particular, overlap of generations and eusociality. While recent systems have relaxed the trait of overlap of generations as a trait for assessment (Costa and Fitzgerald, 1996), they have retained the following criteria for the classification of social systems: cooperative ("alloparental") brood care, reproductive division of labor, parent/offspring aggregation, and aggregation.

Eisenberg (1981; also see Dixson, 1998; Randall *et al.*, 2002) points out that all mammalian sociosexual organizations are variations on three basic themes: the solitary, the monogamous, and the polygamous, although these three systems are reducible to two since solitary organizations are actually polygynous in that a male mates with as many females as possible. The advantage of Eisenberg's (1981) approach, apparent in Figure 8.1, is that variations in sociosexual organization can be relatively straightforwardly related to variations in environmental heterogeneity and differential costs and benefits to female reproductive success. Furthermore, Eisenberg's (1981) system avoids controversies over the precise definitions and importance of "overlap of generations" and "reproductive division of labor," although overlap of generations, one criterion for "social forms *sensu stricto*" as outlined in early classification schemes, is found in all primate species. However, as Costa and Fitzgerald (1996, p. 287) point out in their review, recent schemas for the classification of societies "(1) reflect evolutionary meaningful character traits or phenomena, (2) conceptually unify taxa, and (3) eliminate terminological ambiguity"—which Eisenberg's system fails to do.

The classification systems discussed in Costa and Fitzgerald's (1996) review article that are most amenable to evaluation of mammalian, including primate, sociosexual organization, are those by Gadagkar (1994, 1997) and Sherman *et al.* (1995). Neither of these schemas requires that eusociality be defined as "genetic" division of labor or caste determination (Helms Cahan and Keller, 2003). Thus, primate sociosexual organizations—none of which exhibits genetic division of labor or (nonreversible, morphological) castes—have the potential to reach high grades of sociality within the formulations of Gadagkar (1994, 1997) and Sherman *et al.* (1995). Gadagkar (1994,

Table 8.2. Sociosexual Classification Based upon Gadagkar (1994, 1997), Including Communal, Subsocial, Quasisocial, and Eusocial Structures as well as Definitions, Criteria, and Examples from the Primate Order.

Category	Definition/criteria	Example(s) from the primates
Communal	Members of same generation cooperate in nest building but not in offspring care; no cooperative offspring care, no reproductive division of labor, no parent/offspring aggregation	*Aotus*
Quasisocial/ Subsocial	Adults (usually the mother) care for own offspring for extended period of time; no cooperative offspring care and no reproductive division of labor; aggregation (Wilson, 1971)	*Galago, Otolemur; Microcebus*; some anthropoids?
Eusocial	Members of same generation use same nest or sleeping site; cooperative breeding, reproductive division of labor; aggregation	Some marmosets and tamarins; *Alouatta palliata* (Jones, 1996a; Jones and Agoramoorthy, 2003); *Semnopithecus entellus* (see Hrdy and Hrdy, 1976); some other anthropoids (e.g., *Pan paniscus*; see Vervaecke *et al.*, 2003); *Homo sapiens*

NB: As indicated by Figure 8.1, behavior is expected to be most flexible in multimale–multifemale, cooperative, and polyandrous societies, some of which will be classified as eusocial by criteria employed in the schemas discussed by Costa and Fitzgerald (1996).

1997) emphasizes the presence or absence of altruism in a reproductive context (e.g., cooperative breeding) as a predominant criterion for eusociality. In Gadagkar's (1994, 1997) schema, castes are subordinate to altruism in a general context whose presence or absence distinguishes advanced societies (presence of castes) from primitive (absence of castes) eusocial societies. Cooperatively breeding marmosets and tamarins as well, perhaps, as some other primate taxa demonstrating altruistic reproduction (e.g., alloparenting or, better, alloparenting with nonreproductive "roles") may thus, attain eusocial status (see Nicolson, 1987; Hrdy, 1976; Jones, 1986). In increasing order of social complexity, Gadagkar's (1994, 1997) system includes communal, subsocial, quasisocial, and eusocial societies, the first three grades defined as for earlier schemas. Table 8.2 (after Costa and Fitzgerald, 1996, Table 1) displays this categorization, their defining criteria, and examples from the Primate Order.

The schema of Sherman *et al.* (1995) is also potentially useful for the classification of primate societies since it also emphasizes altruistic reproduction (in particular, cooperative breeding), which they posit is a continuum of traits (the "eusocial continuum") across species. Further, these authors propose that grades of sociality may be measured by an "index of reproductive skew" (see Hager, 2003; Abbott *et al.*, 1998). In this system, only the cooperatively breeding marmosets and tamarins would be classified as eusocial among the primates. One of the unresolved questions for some of the classification schemas discussed by Costa and Fitzgerald (1996) is whether or not reproductive division of labor can be *reversible* (West-Eberhard, 1979). Reversible, facultative, nonobligate reproductive division of labor would incorporate many more primate species within high grades of sociality as formulated in these schemas (e.g., female *A. palliata*; Jones, 1996a; Jones and Agoramoorthy, 2003). Nonetheless, the virtual ubiquity of force, coercion, exploitation, and manipulation in primate, including human, societies is an indication that high grades of sociality (in the West-Eberhardian/Alexandrian sense of the term) are limited within the Order.

A recent paper by Helms Cahan *et al.* (2002) demonstrates the conceptual power of adopting a classification system "inclusive of all social forms." However, it is important to point out that Wcislo (1997) has recently questioned this approach, arguing instead, for treating each taxon as a "phylogenetic entity." To date, primatologists have rarely attempted to integrate their studies of primate societies with those of other taxa (Smuts *et al.*, 1987; Solomon and French, 1997; Jones and Agoramoorthy, 2003), leaving this group of scientists free to make claims about degrees of social complexity for members of the order without regard to criteria in other areas of research on social species (e.g., Boesch *et al.*, 2002). Recent work indicates that, even among social insects, there may be exceptions to generally recognized criteria for social classification schemas (e.g., Queller *et al.*, 2000), suggesting that it is premature to determine how comprehensive general principles of social behavior and organization may be.

Multimale–Multifemale, Cooperative, and Polyandrous Sociosexual Organization and the Expression of Behavioral Flexibility

Figure 8.1 presents a schema suggesting a relationship between extreme environmental heterogeneity and multimale–multifemale, cooperative, and polyandrous sociosexual organization. While all cooperatively breeding marmosets and tamarins would be classified as (primitively) eusocial according to Gadagkar's (1994, 1997) system (e.g., *Saguinus oedipus*, see cover photo), all primate species demonstrating true reproductive altruism (e.g., males in species with female dominance, Jones, 1982b; cercopithecine females in relation to their daughters, Jones, 1983c) would also achieve this grade. Multimale–multifemale, cooperative, and polyandrous sociosexual organizations are of particular interest to primatologists because these associations reflect those of most human societies and because it is within these groups that cooperation between unrelated males and females is sometimes found.

Multimale–multifemale, cooperative, and polyandrous societies are among the most highly gregarious forms of vertebrate social organization, and multimale–multifemale associations occur primarily in the Class Mammalia (Eisenberg, 1981). Emlen and Oring (1977; also see Nunn, 1999, 2003) suggested that multimale–multifemale, cooperative, and polyandrous societies may arise where the environment provides little potential for resource or mate monopolization, in particular, when (1) critical resources are superabundant but widely dispersed; (2) critical resources are sufficiently unpredictable in space and time as to be economically undefendable; and/or, (3) critical resources or mates are clumped in a defensible pattern but the cost of successful defense is too high (due to high population density or resource scarcity). Each of these conditions, or a combination of them, might favor behavioral flexibility in males by favoring genetic polymorphism, genetically-controlled switching, or facultative change to alternative responses for access to females and their gametes (e.g., male–male cooperation, queuing, and other forms of persistence, persuasion, or subordination to females and/or their offspring), and Proulx (1999) provides theoretical support for the conclusion that sexual selection and broad niches (high behavioral flexibility) are positively related.

Related to the topic of sociosexual organization and the potential for behavioral flexibility is Shuster and Wade's (2003) classification of mating systems stressing the operational sex ratio (Emlen and Oring, 1977), in particular, the spatial and temporal distribution of *receptive* females. In this view, females may be a *patchy* resource for males and, consistent with the arguments presented in this volume, the nature and degree (frequency, rate, duration, intensity, and/or quality) of temporal and spatial heterogeneity of receptive females is expected to favor behavioral flexibility in males. As argued above, behavioral flexibility in females is expected to be most sensitive to the temporal and spatial heterogeneity of limiting resources, in particular, food and breeding sites, required for successful reproduction (Chapter 5). As predicted in Figure 8.1, these pressures favoring behavioral flexibility are most likely to be associated with multimale–multifemale, cooperative, and/or polyandrous associations.

It has been suggested that features of populations are expected to be emergent properties of decisions that individuals make in relation to temporal and spatial heterogeneity of their environments, combined with the constraints imposed by phylogeny. This view implies that temporal and spatial patterns at several levels of organization—both abiotic (e.g., climate) and biotic [e.g., intraspecific interactions (frequency, duration, rate, intensity, quality)]—may impose pressure upon the individual, favoring or disfavoring behavioral flexibility. As argued above, the precise optimal response will be a function of the frequency, rate, duration, intensity, and/or quality of perturbation relative to T. As pointed out in Chapter 1, one benefit of sociosexual assemblages is that individuals may be partially *buffered* from environmental

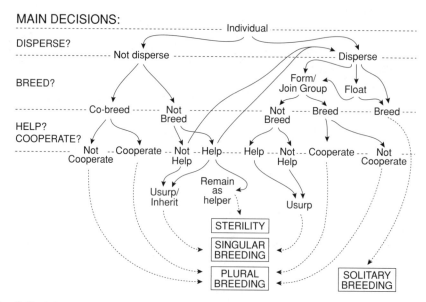

Fig. 8.2. Schematic representation of the range of social decisions available during an individual's lifetime (Helms Cahan *et al.*, 2002, Figure 1, p. 208). Each path of decisions defines a "social trajectory," and relevant theoretical and empirical work has shown that "cooperation among individuals is necessary for evolutionary transitions to higher levels of biological organization" (Velicer and Yu, 2003, p. 75). Solid arrows indicate the order of decisions over time. Dotted lines connect trajectories with the breeding systems (terms in boxes) in which individuals displaying those trajectories may participate. Studies of mammals and social insects have focused upon whether or not an individual attempts to breed or not breed within a group (Keller and Reeve, 1994; Lacey and Sherman, 1997; Hager, 2003). Figure and (modified) legend used with permission.

perturbations, increasing the predictability of his/her spaces, and research on baboons (*P. cynocephalus*) by Bronikowski and Altmann (1996, p. 11) suggests that "social groups rather than populations may be the appropriate unit of analysis for understanding the behavioral ecology of baboons and other highly social primates," an unconventional suggestion in need of theoretical (mathematical) and empirical support.

Conclusions

Future investigations of sociosexual organization and behavioral flexibility in primates also need to evaluate the particular factors and contexts associated with the capacity for switching from one action pattern to another, the developmental trajectory and lability of these behaviors, and the proximate and ultimate costs and benefits of the alternate responses. Such studies will require field manipulations in addition to assessment of maternal and paternal effects. Related research programs might also initiate a new field of behavioral ecological economics to measure and test the decision-making processes of social organisms in the field and in captive conditions, the determinants, costs, and benefits of observed behavioral switches, and the implications of these decisions for survival as well as genotypic and phenotypic success. These investigations would also be helpful in decisions regarding the social grades or trajectories (Fig. 8.2) occupied by primates and how best to define "sociality," in particular, complex sociality (social systems with high reproductive skew), relative to other social species.

The research programs investigating behavioral flexibility might be capable of revealing which heterogeneous components of the abiotic and biotic, including social, environments are most likely to lead to switching by individuals from one response pattern to another and these responses' differential costs and benefits to survival, lifetime reproductive success, and the phenotype. Related to this, if sociosexual assemblages do "buffer" individuals from temporal and spatial environmental heterogeneity, it would be expected that behavioral flexibility would vary with demographic features of populations, in particular, group size and population density. While the particular relationships between behavioral flexibility and sociosexual organization have not been thoroughly analyzed, it is expected that multimale–multifemale, cooperative, and/or polyandrous assemblages will favor behavioral switching, adjusting responses to changing local conditions. One of the reasons that a description of these factors and their interrelationships is important to primatologists and other animal behaviorists is that an understanding of human behavior, perhaps the most flexible extant repertoire, requires knowledge about the differential benefits and costs of facultative responses as well as investment in offspring quality over quantity for the optimization of fitness and phenotypic success. Both of these human traits may be adaptations to environmental heterogeneity for certain individuals in populations or subpopulations with particular features yet to be determined.

Behavioral Flexibility: 9
Interpretations and Prospects

[No] instinct can be shown to have been produced for the good of other animals, though animals take advantage of the instincts of others.

Darwin (1859, p. 208)

Introduction

Phenotypic plasticity represents sets of mechanisms, including behavioral flexibility, of adjustment and adaptation to local conditions that may be components of ontogenetic processes (West-Eberhard, 2003). Slobodkin (1968) and others (Miller, 1956; Slobodkin and Rapoport, 1974; Lerner, 1970; Hochachka and Somero, 1973; Lande, 1980) have proposed a *predictive theory* whereby natural selection favors different mechanisms within and between species for accomodation to local environmental (abiotic and biotic, including social) variations. These authors suggest that the mechanism(s) favored will be a function of the temporal and/or spatial patterning of environmental changes (e.g., fluctuations in food supply, temperature, or rainfall) relative to the organism's generation time (T, "environmental grain"). The proposed mechanism(s) may differ with respect to their rates of activation (i.e., relatively flexible to relatively inflexible) and their sensitivities to local conditions (i.e., relatively attentive to relatively inattentive, see Chapter 2).

These mechanisms have been outlined as the hierarchically arranged, structurally and functionally interactive (a) behavioral; (b) physiological; (c) developmental; and (d) genetic levels (Slobodkin and Rapoport, 1974). *Regulatory behaviors* (e.g., thermoregulatory nesting, classical conditioning, dispersal) are quickly activated by exogenous stimuli and efficiently reversed, providing rapid accomodation to intraseasonal (proximate) conditions. *Physiological compensation* for seasonal environmental fluctuations (e.g., facultative adjust-

123

ment of estrus stage length, acclimatizing ectothermy, pharmacodynamic tolerance) is generally a slower and more durable process than compensation through behavioral regulation.

Developmental processes [e.g., epigenetic processes governing phenotypic expression, rates of cellular organization of anatomy and physiology, postnatal neurogenesis (Weaver *et al.*, 2004; West-Eberhard, 2003; Wourms, 1972)] and, to a greater degree, *genetic differences* for any character or set of characters (West-Eberhard, 2003) represent an investment in a particular phenotypic repertoire that is not reversible within the lifetime of the individual (Lewontin, 1957, 1970; Lerner, 1970). Consistent with Bradshaw's (1965, p. 126) reasoning that, "If a population is subject to recurrent changes in its environment whose duration is the same or less than its generation time, it cannot easily respond to the contrasting environments by directly adaptive genetic changes," the predictive theory proposes that selection pressures for population differentiation in genetically correlated heritable behavioral, physiological, or developmental traits should be weak when the activation of facultative mechanisms compensate for environmental variation. Alternatively, natural selection should favor genetic differentiation of traits where *regulatory responses* cannot compensate for the temporal and spatial patterning of environmental heterogeneity.

The Predictive Theory and Environmental Heterogeneity

These predicted relationships between environmental heterogeneity and epigenetic compensatory responses (e.g., conditioning, cognitive mechanisms) will depend upon the differential mortality and reproduction correlated with alternative responses (Stearns *et al.*, 1991; also see Stearns, 1992) and will be constrained by rates of mutation, recombination, gene flow, and heritability. These predictions have significant implications for the genotypic and phenotypic structures of populations. In particular, where compensatory responses to environmental fluctuations are favored, individuals will be sensitive or attentive to exogenous stimuli, and their response repertoires, consequently, will be governed more immediately by exogenous factors (e.g., humans). Lerner (1970) speaks of "social," "psychological," and "ecological" feedback mechanisms, types of homeostatic responses that are of particular interest where phenotypic responses [see West-Eberhard's (2003) concept of phenotype as "bridge"] are genetically uncorrelated. The predictive theory suggests that the phenotypic variability within populations employing compensatory, or genetically uncorrelated, behavioral, physiological, or developmental responses to environmental fluctuations will be significantly greater than the phenotypic variability within populations or species employing canalized controls in fluctuating environments (see Stearns, 2002).

There is no necessary or sufficient reason to assume an adaptive or genetically correlated basis for continuously distributed characters, and gene transcription does not necessarily accompany the expression of continuous traits (e.g., traits associated with feeding, natality, fertility, reproductive success and rate, or dispersal; Lewontin, 1974; Sammeta and Levins, 1970). Whether or not an action pattern is heritable, however, is of secondary concern to the comparative behaviorist since any response which is correlated with greater relative reproduction or survivorship will promote those genes correlated with the response whatever those genes might regulate or control (West-Eberhard, 1989, 2003). This mode of compensation to local conditions, as discussed in Chapters 1 and 2, is expected to be most beneficial where environmental changes are not trackable spatiotemporally within the physiological limit of the organism (Roughgarden, 1979). Lability of behavior and other responses, however, is expected to be beneficial to fitness only within certain limits beyond which feedback mechanisms break down. There may be a limit to how flexible a suite of behavioral responses can be, in part because of the physiological and genetic loads imposed by coordinating and controlling many neuromuscular options (Relyea, 2002).

The Predictive Theory, Norms of Reaction, and Behavioral Flexibility

Unless complex behavior(s) (e.g., some types of social behavior) is to appear *de novo* (e.g., by "trial and error"), by associative or cognitive processes, or facultatively, it must originate as a result of genetic variation(s) underlying action patterns in response to selection (Page and Erber, 2002). Behavioral flexibility arises in response to differential (genetic or other) costs and benefits over space and time, and individuals within populations whose phenotypes (e.g., morphology, development, physiology, behavior) exhibit a better "fit" to the environments in which they occur will have a higher probability, on average, of surviving and reproducing. The heritable component of the phenotype (i.e., the additive genetic variance) may be transmitted genetically to offspring although the same genotype may have different relative fitnesses in different environments, and different genotypes may have the same average phenotype success ("norms of reaction"; Schlichting and Pigliucci, 1998). This "natural selection" of differentially reproducing genotypes may partially account for changes in frequencies of genotypes from generation to generation within and between populations of a species if selection intensity is strong enough to outweigh other effects (e.g., drift, parasitism, disease; see Gaines and Whittam, 1980).

Stochastic (e.g., founder effects) and other nonselective events (e.g., "random" predation) may also partially determine some component of changes in phenotypes and in the frequencies of genotypes in populations. Some students of organismic and evolutionary biology emphasize the roles of random genetic drift, founder effects, predation, parasitism, disease, habitat mosaics, demography, or other "stochastic" factors in the determination of the genotypic and phenotypic architectures of populations (Gaines and Whittam, 1980; Hausfater *et al.*, 1987; Roughgarden, 1983, 1998; Chapman and Balcomb, 1998). These fruitful criticisms of conventional approaches challenge the assumption that patterns of species dispersions result from chronic or acute resource limitations or other nonrandom, density-dependent determinants of survivorship and fecundity and alert all workers to the importance of empirical tests of alternate hypotheses.

Students of behavior and social organization, including behavioral flexibility, are concerned with the study of *polymorphisms* (genotypically regulated alternative responses), *polyphenisms* (environmentally switched alternatives), *polyethisms* (behavioral alternatives), and *facultative responses* (condition-dependent, nongenetically-programmed alternative responses) within or between individuals and populations over time and space. These researchers may also be interested to describe and to explain the mechanisms and consequences of phenotypic differentiation among individuals, populations, and species that predict the origin of different species assemblages (Kanthaswamy and Smith, 2002; Gavrilets, 2000; Gavrilets and Waxman, 2002). Behavior may be the most important component of the phenotype, and, therefore, the most significant proximate mechanism of population differentiation since the behavioral repertoire represents that component exposed most intimately to the physical and biotic environment (Mayr, 1963; Huey *et al.*, 2003; Altmann *et al.*, 1996). While the translation of evolutionary events into precise genetic terms is the domain of population genetics, it is important for students of behavior and social organization to understand phenotypic models for evolution (Emlen, 1973; Roughgarden, 1979, 1998) in order to evaluate theoretically how a given trait might *behave* in the absence of selection (Roughgarden, 1979) and to sample genotypes within and between study populations.

While this volume primarily addresses the significance of *facultative* behavioral responses, a related concern exists where an association occurs between variations in genotype and continuous traits since there is no necessary *direct* association between them (Lewontin, 1974). Indeed, available evidence shows that physiological, developmental, behavioral, environmental, or some combination of these factors generally mediate the expression of all responses (Pigliucci, 2001; West-Eberhard, 2003). These concerns require the investigator to discriminate four separate though interrelated questions about populations and the individuals from which they emerge. First, to what extent are genetic mechanisms (selective or nonselective) the causes or consequences of population structures? Second, to what extent are variations in genotype, gene transcription, and biochemical efficiency correlated with variations in inclusive fitness within populations? Third, although the amount of genetic variability documented in mammals is relatively modest compared to other vertebrate classes, what is the biological significance of genetic variability within and among populations, in particular, primate populations? Fourth, if populations are not tested under *stress* (Sammeta and Levins, 1970), might research be most likely to detect "levels of chaos in population processes?" (Lidicker, 1981, p. 319).

As Lidicker (1981, p. 319) goes on to say, "Rapidly accumulating information is forcing us to admit, at the point of despair for some, that stochastic processes can be critically important in explaining the biology of populations. We can be reassured, however, that the biosphere is indeed organized; there is negentropy, or life would not be possible. The problem is finding, documenting, and understanding this organization in spite of a great deal of stochastic 'noise'." It should, then, be possible to investigate empirically the causes and consequences of behavioral flexibility, including the underlying genetic variability, if any, of facultative responses (W. Lidicker, personal communication). Methodological problems associated with the statistical analysis of multifactorial variation are inadequately resolved for relatively long-lived organisms that mate nonrandomly, such as primates, particularly if they exhibit *plastic* and/or behaviorally flexible phenotypes (Mather and Jinks, 1982; Smith and Joule, 1981). However, research on mammalian, including nonhuman primate and human, population genetics is vigorous, and a dynamic picture of some relationships is beginning to emerge (Gagneux, 2002).

This volume discusses phenotypic traits within and/or between populations that may be relatively variable (i.e., "plastic," "flexible," and/or "attentive" to the environment). The ideas expressed here also indicate that in some environmental mosaics, behavior(s) exposed to the environment may vary and may be favored *without* underlying genetic correlation(s). It is the latter set of conditions which interests many students of vertebrate behavior since certain learned and cultural processes may represent epigenetic compensatory mechanisms resulting in selection upon hosts, or their prey, or social parasites. While behavioral and population geneticists will labor to demonstrate the genetic and environmental variability underlying phenotypic traits (Gagneux, 2002), students of behavior and social organization may confidently describe the relationship between phenotypic traits and vital parameters (mortality, reproduction) without precise understanding of genetic variance and its phenotypic correlates (Slatkin, 1974; Lande and Arnold, 1983; Robinson, 1999, 2002; Robinson and Ben-Shahar, 2002).

In most instances, the comparative behaviorist's efforts will map distributions of phenotypes with little or no knowledge of allele or genotype distributions but with the assumption that comparisons and contrasts of phenotypic architectures within and between populations reflect environmental patterning by way of rules governing "norms of reaction," the transformation functions converting environmental distributions into phenotypic distributions, and/or "epigenetic programming" by environmental stimuli. New procedures (see Lim *et al.*, 2004; Weaver *et al.*, 2004) should soon permit the comparative behaviorist to assess the extent to which responses are controlled by a genetically induced "switch" mechanism as proposed by West-Eberhard (1979; also see Gross, 1996; Jones and Agoramoorthy, 2003). Dausmann *et al.* (2004), for example, recently reported hibernation in the Malagasy lemurid, *Cheirogaleus medius*, a response likely to be induced genetically and positioned for genomic investigation. As pointed out in the Preface to this volume, however, all responses, whether genetically correlated or not, can be assessed for their possible genetic and/or phenotypic influence(s) on the actor and all individuals influenced by the response(s).

The primatologist (including students of humans) will most often be concerned to describe and to explain behavioral flexibility in characteristics related to interindividual interactions since most primates are obligately social. While the foregoing chapters have attempted to clarify a number of topics related to the evolution of behavioral flexibility in primates, numerous questions remain about which little is known in prosimians, monkeys, apes, and humans). The following sections propose questions requiring attention in future research programs in order to determine the similarities and differences between flexible behaviors within and between species of primates and those of other taxa. Recent studies, for example, raise important questions regarding the grades of sociality within primates, in particular, the definition, features, causes, and consequences of "cooperative breeding" (Clutton-Brock *et al.*, 2003). Another area of study involves the ecological dominance of human beings whose negative consequences upon all major ecosystems impels the student of behavioral flexibility to address the role of individuality and facultative cooperation in the behavioral repertoire of this species (Bolnick *et al.*, 2003). A study of behavioral flexibility is, in great part, a study of the differential costs and benefits of accurate decision-making, another topic discussed below. Finally, the topic of behavioral flexibility is fundamental to an understanding of the worldwide biodiversity crisis, and recent theoretical and empirical work will be summarized in an attempt to identify those topics in greatest need of study. Each of these topics will be addressed in an attempt to highlight their significance for investigations of behavioral flexibility in primates.

What Factors Constrain the Evolution of "True" Social Behavior in Primates?

As discussed in Chapter 8, the level of sociality assigned to primate species depends, in part, upon the classification system utilized. While there is general agreement that no primate species is truly eusocial, demonstrating genetically induced, nonreversible, morphological and nonreproductive castes, recent schemas have broadened the concept of eusociality to include all taxa demonstrating alloparental care. What is unclear about these recent schemas

Fig. 9.1. The Tana mangabey (*Cercocebus galeritus*) depicted here is among the 25 most endangered primate species according to Conservation International (see Wieczkowski, 2003). Behavioral flexibility in this species may be elicited by habitat disturbance. However, it is likely that behavioral flexibility is selected against above some threshold value of disturbance. ©Julie Ann Wieczkowski.

is whether or not the alloparent need himself or (more typically) herself exhibit reproductive restraint (i.e., exhibit temporary or permanent nonbreeding), a requirement of earlier classification systems. Further, the extent to which reproductive restraint need be imposed by a dominant breeder or self-imposed is, likewise, unresolved. What seems clear from theoretical (Doebeli *et al.*, 1997) and empirical (Lacey and Sherman, 1997) studies is that high coefficients of inbreeding combined with stochastic population effects appear to be generally correlated with the evolution of cooperation, including cooperative breeding (Russell and Hatchwell, 2001; Griffin and West, 2003) and eusociality (Doebeli *et al.*, 1997), although kinship does not appear to be a sufficient factor for the evolution and/or maintenance of cooperation (Crespi and Choe, 1997b, pp. 1–2; Dugatkin, 1997; Giraud *et al.*, 2002; see Fig. 9.1). Crespi and Choe (1997b) conclude that genetic, ecological, and phenotypic factors in addition to life history and demography determine individual decision-making, stressing, in particular, the significance of predation and parasitism, constraints on independent breeding, and characteristics of resources required to successfully rear offspring.

Frank (1995) showed that self-restraint is not sufficient to explain the evolution of cooperation. Mutual policing and enforcement of reproductive fairness ("social inhibition"; Beshers *et al.*, 2001; Saltzman, 2003) are necessary conditions for social evolution, including increased social complexity (see, however, Avilés *et al.*, 2002). Recent work with humans supports Frank's theoretical formulations (Bowles and Gintis, 2003; Semmann *et al.*, 2003; Panchanathan and Boyd, 2003), and, in addition to self-restraint, cooperation may be imposed upon subordinates by dominants (see, for example, Hager, 2003a,b; Saltzman, 2003). In addition to policing and punishment, however, other researchers have shown that cooperative behavior in humans and, possibly, some other species, depends upon "trust" and "reputation" as long as the benefits to individuals choosing to cooperate on the basis of these criteria outweigh benefits to individuals helping the untrustworthy (Mohtashemi and Mui, 2003). When assessing the evolution of cooperation and altruism in primates in relation to the mainstream literature on social evolution, it is important to be aware that there is a *disconnect* between insect research assuming that the highest grades of sociality require extreme reproductive skew and primate research generally implying that high grades of sociality involve more or less egalitarian interactions among conspecifics (see, for example, Silk, 2003).

Routes to Eusociality in Primates and Other Species

It is unclear whether any primate exhibits "functional specialization" in cooperative behavior incipient to that found in the castes of eusocial species [e.g., social insects (Wilson, 1971) or naked mole-rats (Lacey and Sherman, 1997)]. Jones (1996a) provided evidence suggesting that female mantled howler monkeys (*Alouatta palliata*) demonstrate temporal division of labor (age-related polyethism) whereby apparently altruistic behavior ("social foraging") is increasingly likely to be exhibited with increasing age and decreased likelihood of selfish reproduction (Chapter 5). Studying meerkats (*Suricata suricatta*), Clutton-Brock *et al.* (2003) concluded that age-related polyethism in this species does not represent "incipient" division of labor in the sense of "functional specialization" found in eusocial taxa. The critical finding by these authors was that meerkats did not specialize in particular activities but that different activities were expressed as a function of age and that "individual differences in foraging success became the principal factor affecting contributions to cooperative behavior" (p. 531), as for female mantled howlers.

The combined results from these studies suggest that factors, possibly energetic ones (Jones and Agoramoorthy, 2003, pp. 124–125; see Russell *et al.*, 2003), related to foraging may be the initial route to advanced sociality in social mammals and, perhaps, other social species as well (Beshers *et al.*, 2001; Wahl, 2002). Crespi and Choe (1997a; also see Crespi and Yanega, 1995) point out the important relationship between eusociality and loss of the capacity to become a reproductive dominant. Cases of incipient primate eusociality, then, should be sought in situations where individuals obtain a low likelihood of achieving reproductive dominance *combined with high reproductive skew* (Crespi and Choe, 1997a, p. 508; Hager, 2003a,b; Saltzman, 2003). Further research is required to document this possibility and to test the claim by Clutton-Brock *et al.* (2003) that age (or size) polyethism is not "incipient" eusociality (functional specialization). Perhaps polyethisms are one route to eusociality (see, for example, Traniello and Rosengaus, 1997; Tofilski, 2002; Keller, 2003; "trajectories", Fig. 8.1).

Traniello and Rosengaus (1997, p. 209), for example, discuss the "dynamic nature and fluidity of task allocation" in social insects and quote Hölldobler and Wilson (1990): "Each species has its own distinctive pattern of temporal polyethism." These authors point out that the study of temporal (age-related) polyethism and related social features encompass the domains of behavioral flexibility, including development. Consistent with the socioecological model advanced by Sterck *et al.* (1997) for primates, Traniello and Rosengaus (1997) emphasize the importance of historical and ecological factors, particularly "nutritional ecology," to explain interspecific variation in social structure. Like primates, some social insects do not display irreversible nonreproductive roles (see Traniello and Rosengaus, 1997, pp. 210–211), exhibiting what these authors term "reproductive plasticity."

Thus, contrary to Clutton-Brock *et al.* (2003), size or age polyethisms may represent "incipient" eusociality, requiring only an evolutionary trajectory toward the loss of reproductive alternatives by (presumably) the subordinate class. Adult mantled howler monkey females, belonging to a species in which age correlates negatively with dominance rank (Jones, 1978, 1980), may represent a case of incipient eusociality among primates (Jones, 1996a; Jones and Agoramoorthy, 2003; see Chapter 5). Interestingly, Crespi and Choe (1997b) point out that high rates of adult mortality can favor eusociality, a relationship in need of investigation for primates and other social vertebrates in order to detect incipient eusociality in these species. In addition to further empirical studies on the behavior and social organization (including group formation; Lacey and Sherman, 1997, pp. 272–276) of this and other species of social mammals demonstrating age or size polyethism, investigations of sociogenomics, ecology, phenotype (including, development and physiology), as well as life history and demography, are required (Crespi and Choe, 1997a,b). It is also of fundamental importance to study the energetics of social relationships. For example, females, "energy-maximizers," may be more social than males (see Wittenberger, 1980) because of the energetic benefits of sociality (see Heinze and Keller, 2000). Females, then, may "drive" social evolution in primates and other social mammals (see Wittenberger, 1980; Lindenfors *et al.*, 2003) because of their fundamental relationship to energetic effects. Males ("time-minimizers") may adjust to female group size (Lindenfors *et al.*, 2003; Wittenberger, 1980) because it is a prudent decision from the standpoint of temporal constraints.

To what Degree Does the Influence of "Individuality" Constrain the Evolution of Sociality in Humans?

Humans and numerous species of insects (e.g., *Apis mellifera*) are highly successful ecologically because of the evolution of sociality in these groups (Keller, 2003; Lewontin, 2000; Jones and Agoramoorthy, 2003). An important task confronting students of complex interindividual behavior, including behavioral flexibility, concerns the description and explanation of similarities and differences in the mechanisms and functions accounting for variations in observed patterns within and between species. Chapter 3 suggests, for example, that the extensive ecological success of humans is more a function of certain individuals (usually, but not necessarily, dominants) exploiting others' (usually, but not necessarily, subordinates') action patterns, resources, thoughts, emotions, and/or relationships, influencing in the process, the genetic, physiological, and developmental pathways related to these responses for selfish advantage.

Humans and most other primates (and mammals) are characterized by relatively small size of reproductive units, generalized ("plastic") phenotypes, individual recognition, and a relatively small range of reproductive differences compared to social insects occurring in large nests with a relatively broad range of functionally specialized phenotypes (see Lacey and Sherman, 1997, pp. 288–289 and 294–295; Kitchen and Packer, 1999; Maynard Smith, 1999). In spite of these distinctions, however, the ecological pressures thought to favor the evolution of sociality in the two groups of organisms are the same [e.g., benefits from predator defense, access to resources (e.g., food, mates), and increased competitive ability], and numerous authors have emphasized the potential value in comparing the evolution of social behavior in vertebrates and invertebrates (Wilson, 1971, 1975; Lacey and Sherman, 1997 and references therein; Robinson *et al.*, 1997; Jones, 1980, 1997a; Abbott *et al.*, 1998; Robinson, 2002; Jones and Agoramoorthy, 2003 and references therein; also see Chapter 8; for a comparison of birds and mammals, see Emlen, 1991). Such an approach privileges an emphasis upon function(s) over an emphasis upon mechanism(s).

Some mechanisms thought to favor the evolution of sociality are, however, hypothesized to be the same for both groups (e.g., kin selection, West-Eberhard, 1975; parental manipulation, Trivers, 1974), although mechanisms favoring cooperation between nonkin (e.g., reciprocal altruism, Trivers, 1971; manipulation by harassment, Stevens, 2004) are expected to be especially important for many primates and certain other social mammals (e.g., dolphins; Connor *et al.*, 2001). Highlighting the importance of high levels of relatedness for social evolution, Chapman *et al.*, 2000 conducted a comparative analysis of the relationship between inbreeding and eusociality in five species of gall-inducing thrips, suggesting that inbreeding is required for the evolution of "bisexual helping" (Lacey and Sherman, 1997). A similar analysis for primates may reveal important relationships between relatedness and levels of sociality, including the transition to associations between unrelated individuals (see Mitani and Watts, 1997).

What Kinds of Social Behaviors Do Humans Exhibit?

It would seem that humans display a combination of discrete traits [e.g., temperament and personality traits cum "individual specialization" (Bolnick *et al.*, 2003)] and more generalized characteristics compounded with the ability to combine and recombine these features in response to past and immediate experience as well as in relation to assessment of costs and benefits for the short, moderate, and long terms. This potential complexity presents a crisis for the expression and elaboration of cooperation and altruism since the benefits of social parasitism (e.g., phenotypic manipulation) and other behaviors subverting an individual's selfish decisions and a conspecific's tendency to assist another's reproduction must be high (Bronstein, 2001, 2003).

A major theme of Chapter 3 and the present section is that, for humans and, perhaps, other social mammals, tactics and strategies of interindividual exploitation and manipulation drive the expression of behavioral flexibility. Indeed, as suggested by Crook (1970), ostensible cooperation in primate and other mammalian societies is a "subterfuge" (p. xxix) for selfish ends designed to control others. Crook implies that, in mammalian societies, this condition is derived from what Wheeler (1934, 1939 quoted in Crook, 1970, p. xxix) called, "the problem of the male" who can be considered a (social) parasite upon females (also see Smuts, 1985, 1987b; Alexander *et al.*, 1997; West-Eberhard, 2003, pp. 630–637; and, Jones and Agoramoorthy, 2003, pp. 113–122 on the problems males pose for females). Future research [e.g., sequence analysis of likelihoods of escalation of interactions (Jones, 1983a, Figure 3), duration of relationships, and techniques of policing and punishment] is required to determine the extent to which sociality in humans and other primates entails assisting the reproduction of conspecifics. Phylogenetic analyses and studies of "sociogenomics" may reveal important evolutionary transitions in primate evolution, including similarities and differences between Neotropical and Paleotropical primate taxa.

How Important is the Accuracy of a Flexible Behavioral Response?

It is necessary to assume that, on average, behavioral flexibility in heterogeneous regimes has been functionally adaptive (Sultan and Spencer, 2002; Miller, 1997), and numerous factors will influence an individual's decisions (conscious or otherwise) to behave in one way or another. Behavioral flexibility affords a toolbox of potential responses over time and space. In general, the response, or set of responses, selected is expected to be a function of the response's fitness value (including the actor's expected lifespan), its degree of genetic programming, its temporal and energetic costs, previous experience with the consequences of the response, the degree to which the individual's responses have been shaped by one or more social parasite, and the extent to which the decision to act is based upon the expectation of immediate or future gains. Differential choice of response(s) is also expected to be a function of the availability of information (Panchanathan and Boyd, 2003) since the individual's decision(s) will be a function of his/her sensation and perception of environmental stimuli. As pointed out in Chapter 1, one goal of the study of behavioral flexibility is to understand how organisms manage and minimize uncertainty and risk, and "the organism's system of perception and response" (Sultan and Spencer, 2002, p. 272) will in great part determine his/her response accuracy for enhancement of environmental predictability and security.

Sultan and Spencer (2002; also see Crowley, 2003) discuss those conditions influencing response accuracy. Since response accuracy will, in part, depend upon the predictability inherent to environmental heterogeneity (e.g., cycles of rainfall; Jones, 1997b; rates of interaction with other organisms, especially conspecifics), response accuracy, and the evolution of plasticity can be negatively impacted by cues and signals that are unreliable. As pointed out by Sultan and Spencer (2002), such constraints entail a circumstance whereby the developmental and selective environments are decoupled, yielding no or poor correlation between the two. In these conditions, the individual cannot "track" the environment to his/her own advantage. Similarly, these authors point out that response accuracy may be constrained in "labile" environments or "when the required phenotypic response entails a long lag time" (p. 272). Again, these constraints yield conditions in which the individual's responses are uncorrelated with (potential) environmental stimuli. On the other hand, Moran (1992) has shown that selection on the organism's sensory, perceptual, or other response systems may improve response accuracy.

Moran's (1992) theoretical work showed an inverse relationship between response accuracy and environmental heterogeneity. Thus, greater environmental heterogeneity lowers the threshold for beneficial expression of flexible and/or plastic behavior. Indeed, in Sultan and Spencer's (2002) model, plasticity will be "fixed" regardless of a significant degree of inaccuracy as long as moderate dispersal rates are found. Further, their model indicates that selection is sensitive to small changes in accuracy since small increases in response accuracy lead to replacement of specialists by plastic phenotypes. The work of Moran (1992) and Sultan and Spencer (2002 and references therein) strongly suggest that, in heterogeneous regimes, selection on sensory, perceptual, memory, learning, and other cognitive mechanisms may have been strong since the potential (fitness) benefits yielded appear to be high. Related to these inferences, it might be expected that organisms utilizing higher-order processes of decision-making are likely to be more vulnerable to costs of inaccuracy since their decisions will often be based upon the expectation of (uncertain) future benefits. Interestingly, Crowley (2003) suggests that the display of behavioral flexibility is not necessarily dependent upon environmental variation where higher-order cognition (e.g., categorization, see Chapter 4) operates. It will be necessary, for researchers to determine the extent to which certain "psychological" characteristics of primates and, possibly, other social vertebrates, might be decoupled from (past and/or immediate) environmental influences.

Behavioral Inefficiency and Behavioral Flexibility

Recent work on humans demonstrates that certain responses are highly "inefficient" (Pelli et al., 2003), and De Jaegher (2003) has suggested that "error-proneness" may function as a handicap signal. Further research is required to assess the differential costs and benefits (tradeoffs) of response accuracy in primates and other taxa related to heterogeneous regimes (Porter and Blaustein, 1989; Hauber and Sherman, 2001). Such studies may involve the manipulation of risk in various contexts (e.g., exposure to predators, foraging, mate selection or guarding) to reveal the potential expression of a range of strategies associated with varying tradeoffs. Bees, for example, trade off foraging speed for accuracy (Chittka et al., 2003), and incubating birds demonstrate flexible behavior in response to variations in predation risk (Ghalambor and Martin, 2002). For flexible decision-making in the social domain (Neff and Sherman, 2002; Alberts, 1999), it is important to assess the recognition mechanisms, rules, and accuracy of choices as these features relate to environmental heterogeneity over time and space (Bergman et al., 2003). For example, in some species (e.g., primates), females cannot store sperm, a condition that may select for mechanisms of (overt or "cryptic") discrimination to reduce their choice of "costly" males (see, for example, Lorch and Chao, 2003; Reeder, 2003). Flexibility of female discrimination mechanisms might be selected in response to dynamic or subtle male traits (e.g., variations in sperm quality, mating rates, age, dominance rank, personality, paternal investment).

Bergman et al. (2003) have recently reported "that baboons recognize that a dominance hierarchy can be subdivided into family groups" (p. 1234). These authors suggest that such capacities for discrimination imposed by social complexity may have favored higher-order cognitive processing and represent a "precursor to some components of human cognition" (p. 1234). Since social parasitism (e.g., phenotypic manipulation) may select for sociality in some regimes (Crespi and Choe, 1997a; Schwarz et al., 1997), one might predict the same sort of discrimination capacities described for baboons by Bergman and his associates in other taxa. It is important to determine why social parasitism and other selection pressures favoring sociality (e.g., predation) have apparently led, in some taxa (e.g., many primate species, including humans) to higher-order cognition and individuality, and in other taxa (e.g., social insects) to alternate discrimination and information-processing mechanisms (e.g., olfaction) and to castes ("true sociality"). Crespi and Choe (1997a) identify the factors that may differentiate these alternative routes ("trajectories," Fig. 8.1) to complex sociality, and Hamilton and Dill (2002) present a preliminary model assessing the role of social parasitism in the evolution of complex social relations.

Toward an Uncertain Future: Behavioral Flexibility and the Conservation of Primate Biodiversity

As discussed in Chapter 2, habitat disturbance may favor the expression of behavioral flexibility and plasticity, for example, by increasing dispersal rates of both sexes. Sultan and Spencer (2002), however, point out that extreme habitat disturbance may result in the loss of plasticity or behavioral flexibility if response accuracy decreases below some threshold value that would be typical to a given species and its tolerance for a range of habitats. Lee (1997), for example, demonstrated a pattern of adjustment to habitat disturbance by Sulawesi crested black macaques (*Macaca nigra*) that increased their risk of extinction. Populations of these monkeys were declining despite demonstrating the ability to modify their ranging and foraging patterns in accord with habitat quality. Lee points out that, while most species of macaques demonstrate a noteworthy degree of behavioral flexibility in response to environmental variations, behavioral flexibility in crested black macaques appeared to be limited by increasing their foraging range for few food types rather than increasing their niche breadth, as has been documented for other species of macaques. Lee also documented several other features of the behavioral repertoire of *M. nigra* in disturbed habitats that limited their flexibility and capacity for facultative responses to changing conditions (e.g., restricted social repertoires).

Lee's (1997) study and others (see, for example, Mitchell, 1994; Wieczkowski, 2003) suggest that the ability to increase niche breadth is a critical factor in coping with deforestation and avoiding extinction of some primate species, a response that will depend not only upon the biology of the species but also upon characteristics of guilds, phenogroups, populations, and ecosystems (e.g., coefficients of competition). On the other hand, some species may increase their chances of survival by narrowing niche breadth (increasing their degree of specialization), an option that may be opposed for most primates and many other mammals because of their preadaptations to generalist phenotypes.

Jones (1997c; also see Harcourt *et al.*, 2002) discussed "rarity in primates," identifying several features associated with endangered primate species: narrow geographic ranges, island endemism, preference for primary forest (especially wet or humid forest), and wide niches likely to overlap with resource competitors. It is important for primatologists to test hypotheses designed to assess the capacity of primates to increase or to decrease niche breadth under conditions of habitat disturbance, when it is beneficial for them to do so. It will also be necessary to compare and contrast these measures across a range of environmental conditions, including conditions varying in intensities of competition within and between species. Indeed, it may be beneficial to formalize *the environmental potential for behavioral flexibility* relative to a given primate group or population as a function of the dispersion and quality of its resources and its competitive regime. Such measurements might permit some degree of predictive power in evaluating short-, medium-, and long-term probabilities of viability in local conditions.

The studies by Sultan and Spencer (2002, also see Hanski, 2001) target conditions in which severe habitat destruction weakens or eliminates the correlation between the responses of organisms and environmental cues. These authors' research highlights the critical importance of estimating dispersal rates for primate metapopulations and using the stability, decrease or increase in these rates as partial assays of response to increasing environmental heterogeneity. A decrease or loss of flexibility or plasticity indicated by decreases in dispersal rates or by a failure to increase dispersal rates as habitat disturbance increases (above some unknown threshold) could be a marker for increasing vulnerability to extinction (see Fig. 9.1).

Species might also be made more "extinction prone" by decreases in food quality which some studies suggest is much more important than food abundance as a predictor of species abundance (Wasserman and Chapman, 2003; Estrada *et al.*, 2002; Wieczkowski, 2003 and references therein). Thus, variations in behavioral flexibility (e.g., responses to habitat disturbance) may occur, in part, as a function of variations in the availability of energy, which may, in turn, influence the life history tactics within populations. French (1997), for example, summarizes data showing that variations in gestation costs are correlated with female reproductive tactics (interbirth intervals) for several species of cooperatively breeding callitrichids, and it would seem important to investigate whether or not these variations correspond to species abundance and/or conservation status. French's (1997, Figure 3.5) treatment suggests that the more endangered forms (e.g., *Leontopithecus, Callimico*) experience greater energetic constraints. It has been suggested that some forms of social behavior (e.g., social parasitism) evolved as energy-saving strategies (Heinze and Keller, 2000; Jones and Agoramoorthy, 2003; Lewis and Pusey, 1997), a hypothesis that may apply, as well, to the evolution of several features of cooperatively breeding primate taxa (e.g., paternal care). Behavioral flexibility (e.g., paternal care, twinning), then, may arise in response to heterogeneity in the availability of energy, a relationship in need of further investigation.

Conclusions

The worldwide biodiversity crisis increases the significance of research on all aspects of behavioral flexibility, plasticity, and habitat disturbance. Nonetheless, it is also important to compare and contrast the evolution of sociality in social insects and primates for an understanding of their worldwide ecological success. Despite certain similarities (convergence) in the routes to sociality of these groups (Jones and Agoramoorthy, 2003; Helms Cahan *et al.*, 2002), different "social trajectories" have resulted in a number of differences in the mechanisms employed for ecological dominance by social means (Fig. 8.2; Crespi and Choe, 1997). Currently, primatologists are primarily focused on those mechanisms characteristic of the cognitive capacities of primates, capacities presenting clear benefits (e.g., abstract reasoning) as well as (less studied) costs (e.g., error induced by complexity). Future studies need to resolve the degree to which the behavior of primates and other social mammals depends upon higher-order cognitive processes, and one of the most important domains of future investigation concerns the differential costs and benefits (genetic and other) of higher mental processes relative to environmental conditions. When this research program, including captive and field experiments and studies on "sociogenomics," is more advanced, primatologists and other animal, including human, behaviorists will be positioned to evaluate the role played by behavioral flexibility in the ecological success of species (e.g., humans) and in the preservation of biodiversity. At once, the study of behavioral flexibility has the potential to serve the ends of conservation biology and of investigations into broad principles of behavior and social organization.

Glossary

Selected terms and concepts employed in the present volume for a discussion of behavioral flexibility.

Adaptation: Evolutionary response (→fitness) to environmental conditions. Environments may be simple or complex (multifactorial, including interactive), and adaptation may be local or global.

Age polyethism: See "temporal polyethism."

Behavioral flexibility: Reversible, within-individual alternative behavioral phenotypes.

Categorization: The process of differentiating events in the world into groups using the rule "similar to or different from" (McGarty and Turner, 1992).

Coefficient of competition: A measure of ecological overlap (resource utilization) of members of a community (Schoener, 1974).

Cognition: Higher-order neural processes, not necessarily conscious and aware.

Communication: The provision of information by a sender to a receiver, and the subsequent use of that information by the receiver in deciding how to respond (Bradbury and Vehrencamp, 1998).

Compromise models of reproductive skew: These models assume that neither dominant nor subordinate controls reproduction (see Hager, 2003). "Tug-of-war models" assume that both dominant and subordinate allocate a fraction of group productivity to themselves to increase their share of reproduction.

Condition: The proportion of total available resources acquired by an individual.

Condition-dependent: Relative to or dependent upon individual state, including potential monopolization of limiting resources.

Cue: A stimulus whose perception by other organisms is not beneficial to the emitter (Bradbury and Vehrencamp, 1998).

139

Decision: An endogenously or exogenously induced rule leading to a response. A response may be a decision not to respond.

Division of labor: The apportionment of reproductive labor within a group or society. In the extreme case, some individuals may forego all or some proportion of their selfish reproduction to assist other members of the group or society to raise offspring, and may demonstrate irreversible morphological and behavioral characteristics. Any set of individuals in a given group or society that exhibits both morphological and behavioral specializations (Wilson, 1971, Chapter 11). The presence or absence of division of labor is often employed as a diagnostic criterion for higher grades of sociality (Chapter 8).

Environmental grain: The effective size of a "patch" relative to the size and motility of an organism (Emlen, 1973).

Environmental heterogeneity: Spatial and temporal variation in stimulus or stimulus array endogenous or exogenous to the organism and representing selection pressures possibly resulting in the maintenance of genetic polymorphism.

Epigenetic effects: Within-individual heritable causal interactions between genes and their products during development. Also discussed as contributions by genes in certain cells to genetic effects in other cells.

Facultative response: Response capable of expression under varying environmental conditions. Reversible phenotypic alternative. Facultative responses (e.g., timing of births) are ubiquitous in primates. Nothing is implied about the role of genes in determining the phenotype although the "switch" mechanism is assumed by most authors to be genetically induced (see "polyphenism" below). "Facultative" response is sometimes used interchangeably with "condition-dependent" response.

Female emancipation: Condition operating when females are relatively unencumbered by direct monopolization by males in time and space. Female emancipation is thought to result from high unpredictability and/or heterogeneity of limiting resources required by females (Emlen and Oring, 1977) and may facilitate intersexual selection ("female choice"). Probable examples of "female emancipation" in primates occur in *Alouatta palliata* (Jones and Cortés-Ortiz, 1998), *Brachyteles arachnoides* (Strier, 2000), and *Pan paniscus* (Vervaecke *et al.*, 2003).

Functional magnetic resonance imaging (fMRI): The recording of firing nerve cells, which metabolize oxygen from surrounding blood. This imaging process produces a scan of neural processes not measurable by observing that component of the phenotype exposed to the environment. Interpretation of fMRI scans is controversial.

Generalist: Species or individuals adopting a broad range of solutions (e.g., social parasitism, increased niche breadth) to environmental challenges.

Generation time (T): The average age at which a female reproduces.

Genetic correlation: Association between alleles and traits.

Genomic imprinting: An epigenetic process whereby alleles from both parents are present but only one is expressed due to some mechanism of inactivation (e.g., gametogenic suppression).

"Green beard" effect: An "outlaw" gene inducing the phenotype to employ a trait "as a convenient *label* for the presence or absence of itself" (Dawkins, 1999, p. 143, emphasis in original).

Individuality: The capacity to express one's own selfish interests (directly or indirectly) relative to abiotic and biotic, including social, constraints. A component of "individuality" is those traits characteristic of the individual that are relatively stable over time and space ("personality").

Intergenomic conflict: Conflicts among genes between organisms.

Interlocus contest evolution: An evolutionary "chase" resulting from genomic conflict.

Intersexual antagonistic coevolution ("sexual conflict"): "[T]he evolution of traits that increase the reproductive success of members of one sex at a cost to members of the opposite sex" (Holland and Rice, 1999). Intersexual antagonistic coevolution is a form of intergenomic conflict.

Intersexual ontogenetic conflict: A type of conflict "manifest during development when expression of the same allele, on average, moves one sex toward, and the other sex away from, its phenotypic optimum" (Rice and Chippendale, 2001, p. 685).

Intragenomic conflict: Conflicts among genes within organisms.

Intraspecific social parasitism: The coexistence of conspecifics in which one or more exploits the time and energy of others for fitness benefits to the parasite(s) at a fitness cost to the host(s).

Learning: A relatively permanent change in behavior resulting from experience.

Lifetime reproductive success: Number of offspring an individual produces over a lifetime.

Maternal effects: A component to environmental variation representing covariances between the maternal environment (from cytoplasm to maternal care) and a mother's condition.

Motivation: Mechanisms that initiate, maintain, and direct response(s).

Negative phenogroup assortment: Disfavoring members of one's phenogroup.

Negative reinforcement: Removal of an aversive stimulus leading to an increased likelihood of the operative behavior's expression.

Norm of reaction: The function of the correspondence between different environments and their resulting phenotypes.

Obnoxious phenotype: A phenotype having a disruptive and/or damaging effect on one or more conspecifics' phenotypes by employing offensive or objectionable behavior to obtain selfish benefits.

Operational sex ratio (OSR): The ratio of reproductively active males to fertilizable females.

Persistence: Maintaining one or more states or conditions over time as a tactic to obtain mating or other opportunities potentially benefiting fitness.

Persuasion: Employing one or more behavioral tactics to influence a conspecific's behavior in a manner beneficial to one's own interests.

Phenogroup: A set of individuals exhibiting similar traits for resource exploitation.

Phenotype matching: Employment of one or more of an individual's traits to direct responses to other conspecifics sharing those phenotypic characteristics.

Phenotypic flexibility: The reversible within-individual component of phenotypic plasticity. The focus of these studies in primatology often involves studies of behavior (e.g., behavioral flexibility) or physiology (e.g., acclimation or acclimatization).

Phenotypic plasticity: Phenotypic variation expressed by reproductive individuals throughout their lifetime. Following models of population genetics (after Piersma and Drent, 2003, p. 231), phenotypic plasticity is understood to include the total phenotypic variance of a population divisible into a genotypic (G) component, an environmental (E) component, and interaction between these (G × E). Also included in total phenotypic variance is an error term, e. G is divisible into three effects (additive, dominance, and epistatic), and measures of phenotypic plasticity can vary over time within individuals. Within-individual variation = G + reversible and irreversible fractions of phenotypic variance. As applied to quantitative (continuously distributed) characters, research (especially in behavioral genetics) has often focused upon estimates of heritability (the heritable fraction of total phenotypic plasticity). Recent treatments have advised subdividing E into two measures: a reversible fraction (flexibility) and an irreversible fraction (ontogenetic or developmental stage) of total phenotypic variance. While most authors argue or imply that the phenotype is a function of developmental stage (see Piersma and Drent, 2003, p. 231), West-Eberhard (2003) discusses the phenotype as a "bridge" between the environment and the genotype.

Pleiotropy: Multiple effects of a gene.

Polymorphism: Genotypically regulated alternative responses implying a locus with two or more alleles. For example, cooperatively breeding marmosets and tamarins may be polymorphic for helping behavior, possibly in response to population density (interaction rates) and/or the availability or breeding sites.

Polyphenism: Environmentally switched alternatives. Most discussions assume that the "switch" is "hard-wired."

Polyspecialist: Species or individuals exhibiting a limited number of distinctive alternative characteristics in behavior, morphology, physiology, cognition, etc.

Positive phenogroup assortment: Favoring members of one's phenogroup.

Quantitative character: Genes contributing to continuous (quantitative) variation in a phenotype for which the average phenotypic difference between genotypes are small compared with the variation between individuals within genotypes.

Queuing: Waiting for things, events, states, or conditions.

Reproductive skew: The apportionment of reproduction among same-sex members of a group. At one extreme, one individual monopolizes all breeding (e.g., some marmosets and tamarins), at the other extreme, reproductive output is distributed more or less equally among all individuals of a sex (female *Brachyteles*). All other things being equal, reproductive skew will be higher in males than in females (Trivers, 1972).

Reproductive value (v_x): The relative number of offspring that will be produced by each female surviving to age "x."

Resource dispersion: Distribution and abundance of limiting resources in time and space. It is often necessary, in addition, to discuss *resource quality*.

Resource holding potential (RHP): Ability of an individual to monopolize a limiting resource (e.g., food, mates).

Resource quality: The characteristics of a resource, usually a limiting resource, beneficial to the consumer.

Response accuracy: Percent adaptive or beneficial responses.

Signal: The vehicle providing the information in communication (Bradbury and Vehrencamp, 1998).

Signature: As employed in this book, a trait, schema, or behavioral program characteristic of primates and presumed to be expressed for the optimization of fitness in response to environmental heterogeneity.

Social cognition: The ways in which organisms understand conspecifics as "intentional and mental agents" (Call and Tomasello, 2003; Bering and Povinelli, 2003; Essock-Vitale and Seyfarth, 1987).

Social learning: Learning by observation, particularly observation or imitation of conspecifics.

Specialist: Species or individuals exhibiting a narrow range of traits especially suited to a particular regime (e.g., food supply or social conditions).

Temporal polyethism: The regular changing of labor roles by members of a group or society as they age.

Theory of mind: A state in which one individual is capable of assuming the perspective of a conspecific and assesses the other's differential (condition-dependent) costs and benefits.

Totipotent: The ability of an individual to perform most or all of a society's roles.

Transactional models of reproductive skew: The primary assumption of these models of reproductive skew is that a dominant individual controls group membership and benefits from the presence of subordinates (see Hager, 2003). Transactional models are classified as either "concession models" or "restraint models". In the former, treatments analyze relations between subordinates and dominants in a variety of conditions (e.g., when subordinates are evicted). In the latter, subordinates are assumed to be limited only by the dominant's ability to evict the subordinate from the group.

References

Abbott, D. H. 1993. Social conflict and reproductive suppression in marmoset and tamarin monkeys. In: W. A. Mason, and S. P. Mendoza (eds.), *Primate Social Conflict*, pp. 331–372, State University of New York Press, Albany.

Abbott, D. H., Saltzman, W., Schultz-Darken, N. J., and Tannenbaum, P. L. 1998. Adaptations to subordinate status in female marmoset monkeys. *Comp. Biochem. Physiol.* Part C **119**:261–274.

Alberts, S. C. 1999. Paternal kin discrimination in wild baboons. *Proc. R. Soc. Lond. B* **22**:1501–1506.

Alberts, S. C., Watts, H. E., and Altmann, J. 2003. Queuing and queue-jumping: Long-term patterns of reproductive skew in male savannah baboons, *Papio cynocephalus. Anim. Behav.* **65**:821–840.

Alexander, R. D. 1974. The evolution of social behavior. *Ann. Rev. Ecol. Syst.* **5**:325–384.

Alexander, R. D., Hoogland, J. L., Howard, R. D., Noonan, K. M., and Sherman, P. W. 1979. Sexual dimorphism and breeding systems in pinnipeds, ungulates, primates and humans. In: N. A. Chagnon, and W. Irons (eds.), *Evolutionary Biology and Human Social Behavior: An Anthropological Perspective*, pp. 402–435, Duxbury Press, North Scituate.

Alexander, R. D., Marshall, D. C., and Cooley, J. R. 1997. Evolutionary perspectives on insect mating. In: Choe, J. C., and Crespi, B. J. (eds.), *The Evolution of Mating Systems in Insects and Arachnids.* pp. 4–31, Cambridge University Press, Cambridge.

Altizer, S., Nunn, C. L., Thrall, P. H., Gittleman, J. L., Antonovics, J., Cunningham, A. A., Dobson, A. P., Ezenwa, V., Pedersen, A. B., Poss, M., and Pulliam, J. R. C. 2003. Social organization and parasite risk in mammals: Integrating theory and empirical studies. *Ann. Rev. Ecol. Evol. Syst.* **34**:517–547.

Altmann, J. 1974. Observation study of behavior: sampling methods. *Behaviour* **49**:227–267.

Altmann, J., Altmann, S. A., and Hausfater, G. 1978. Primate infant's effects on mother's future reproduction. *Science* **201**:1028–1030.

Altmann, J., Alberts, S. C., Haines, S. A., Dubach, J., Muruthi, P., Coote, T., Geffen, E., Cheesman, D. J., Mututua, R. S., Saiyalel, S. N., Wayne, R. K., Lacy, R. C., and Bruford, M. W. 1996. Behavior predicts genetic structure in a wild primate group. *Proc. Nat. Acad. Sci. USA* **93**:5797–5801.

Altmann, S. A. 1998. *Foraging For Survival: Yearling Baboons in Africa.* The University of Chicago Press, Chicago.

Amarasekare, P. and Nisbet, R. M. 2001. Spatial heterogeneity, source-sink dynamics, and the local coexistence of competing species. *Am. Nat.* **158**:572–584.

Anderson, C. M. 1992. Male investment under changing conditions among Chacma baboons at Suikerbosrand. *Am. J. Phys. Anthropol.* **87**:479–496.

Anderson, J. R. 1995. *Cognitive Psychology and Its Implications*, 4th edn. W. H. Freeman and Company, New York.

Andersson, M. 1994. *Sexual Selection*. Princeton University Press, Princeton.

Andres, M., Solignac, M., and Perret, M. 2003. Mating system in mouse lemurs: theories and facts, using analysis of paternity. *Folia Primatol.* **74**:355–366.

Andrews, J. 1998. Infanticide by a female black lemur, *Eulemur macaco*, in disturbed habitat on Nosy Be, north-western Madagascar. *Folia Primatol.* **69** (Suppl. 1):14–17.

Arnqvist, G., and Rowe, L. 2002. Antagonistic coevolution between the sexes in a group of insects. *Nature* **415**:787–789.

Aureli, F., and Whiten, A. 2003. Emotions and behavioral flexibility. In: D. Maestripieri (ed.), *Primate Psychol.*, pp. 289–323, Harvard University Press, Cambridge.

Austad, S. N. 1984. A classification of alternative reproductive behaviors and methods for field-testing ESS models. *Am. Zool.* **24**:309–319.

Avilés, L., Abbot, P., and Cutter, A. D. 2002. Population ecology, nonlinear dynamics, and social evolution. I. Associations among nonrelatives. *Am. Nat.* **159**:115–127.

Bagemihl, B. 1999. *Biological Exuberance: Animal Homosexuality and Natural Diversity*. St. Martin's Press, New York.

Baglione, V., Canestrari, D., Marcos, J. M., and Ekman, J. 2003. Kin selection in cooperative alliances of carrion crows. *Science* **300**:1947–1949.

Baldwin, J. D., and Baldwin, J. I. 1976. Vocalizations of howler monkeys (*Alouatta palliata*) in western Panama. *Folia Primatol.* **26**:81–108.

Baldwin, J. M. 1902. Modification. In: J. M. Baldwin (ed.), *Dictionary of Philosophy and Psychology*, p. 94, The Macmillan Company, New York.

Bales, K., O'Herron, M., Baker, A. J., and Dietz, J. M. 2001. Sources of variability in numbers of live births in wild golden lion tamarins (*Leontopithecus rosalia*). *Am. J. Primatol.* **54**:211–221.

Bales, K., Dietz, J., Baker, A., Miller, K., and Tardif, S. D. 2000. Effects of allocaregivers on fitness of infants and parents in callitrichid primates. *Folia Primatol.* **71**:27–38.

Balistreri, M. 2003. *The Evasion-English Dictionary*. Melville House Publishing, Hoboken.

Barclay, J. J., and Gregory, P. 1981. An experimental test of models predicting life history characteristics. *Am. Nat.* **117**:944–961.

Barrett, L., and Henzi, S. P. 2002. Constraints on relationship formation among female primates. *Behaviour* **139**:263–289.

Barrett, L., Gaynor, D., and Henzi, S. P. 2002. A dynamic interaction between aggression and grooming reciprocity among female chacma baboons. *Anim. Behav.* **63**:1047–1053.

Barrett, L., Henzi, S. P., Weingrill, T., Lycett, J. E., and Hill, R. A. 1999. Market forces predict grooming reciprocity in female baboons. *Proc. R. Soc. Lond. B* **266**:665–670.

Barrett, L., Henzi, S. P., Weingrill, T., Lycett, J. E., and Hill, R. A. 2000. Female baboons give as good as they get, but do not raise the stakes. *Anim. Behav.* **59**:763–770.

Barsalou, L. W. 1992. *Cognitive Psychology: An Overview for Cognitive Scientists.* Lawrence Erlbaum, Hillsdale.

Bee, M. A., Perrill, S. A., Owen, P. C. 2000. Male green frogs lower the pitch of acoustic signals in defense of territories: a possible dishonest signal of size? *Behav. Ecol.* **11**:169–177.

Beekman, M., Komdeur, J., and Ratnieks, F. L. W. 2003. Reproductive conflicts in social animals: who has power? *Trends Ecol. Evol.* **18**:277–282.

Bergman, A., and Siegal, M. L. 2003. Evolutionary capacitance as a general feature of complex gene networks. *Nature* **424**:549–552.

Bergman, T. J., Beehner, J. C., Cheney, D. L., and Seyfarth, R. M. 2003. Hierarchical classification by rank and kinship in baboons. *Science* **302**:1234–1236.

Bering, J. M., and Povinelli, D. J. 2003. Comparing cognitive development. In: D. Maestripieri (ed.), *Primate Psychology*, pp. 205–233, Harvard University Press, Cambridge.

Bernstein, I. S., Gordon, T. P., and Rose, R. M. 1974. Aggression and social controls in rhesus monkey (*Macaca mulatta*) groups revealed in group formation studies. *Folia Primatol.* **21**:81–107.

Beshers, S. N., Huang, Z. Y., Oono, Y., and Robinson, G. E. 2001. Social inhibition and the regulation of temporal polyethism in honey bees. *J. Theor. Biol.* **213**:461–479.

Bicca-Marques, J. C. 2003. How do howler monkeys cope with habitat fragmentation? In: L. K. Marsh (ed.), *Primates in Fragments: Ecology and Conservation*, pp. 283–304, Kluwer, New York.

Bjedov, I., Tenaillon, O., Gérard, B., Souza, V., Denamur, E., Radman, M., Taddie, F., and Matic, I. 2003. Stress-induced mutagenesis in bacteria. *Science* **300**:1404–1409.

Boesch, C., Hohmann, G., and Marchant, L. F. (eds.) 2002. *Behavioural Diversity in Chimpanzees and Bonobos.* Cambridge University Press, Cambridge.

Boinski, S. 1999. The social organizations of squirrel monkeys: implications for ecological models of social evolution. *Evol. Anthropol.* **8**:101–112.

Boinski, S., Sughrue, K., Selvaggi, L., Quatrone, R., Henry, M., and Cropp, S. 2002. An expanded test of the ecological model of primate social evolution: competitive regimes and female bonding in three species of squirrel monkeys (*Saimiri oerstedii, S. boliviensis,* and *S. sciureus*). *Behaviour* **139**:227–261.

Bolnick, D. I., Svanbäck, R., Fordyce, J. A., Yang, L. H., Davis, J. M., Hulsey, C. D., and Forister, M. L. 2003. The ecology of individuals: Incidence and implications of individual specialization. *Am. Nat.* **161**:1–28.

Bowles, S., and Gintis, H. 2003. The evolution of cooperation in heterogeneous populations. *Santa Fe Institute Working Paper* #03-05-031 (Abstract).

Box, H. O. (ed.) 1991. *Primate Responses to Environmental Change.* Chapman and Hall, London.

Box, H. O. 2003. Characteristics and propensities of marmosets and tamarins: implications for studies of innovation. In: S. M. Reader, and K. N. Laland (eds.), *Animal Innovation*, pp. 197–219, Oxford University Press, Oxford.

Bradbury, J. W., and Vehrencamp, S. L. 1977. Social organization and foraging in emballonurid bats III: mating systems. *Behav. Ecol. Sociobiol.* **2**:1–17.

Bradbury, J. W., and Vehrencamp, S. L. 1998. *Principles of Animal Communication.* Sinauer Associates, Inc., Sunderland.

Bradshaw, A. D. 1965. Evolutionary significance of phenotypic plasticity in plants. *Adv. Genet.* **13**:115–155.

Brockett, R. C., Horwich, R. H., and Jones, C. B. 2000a. Female dispersal in the Belizean black howling monkey (*Alouatta pigra*). *Neotrop. Primates* **8**:32–34.

Brockett, R. C., Horwich, R. H., and Jones, C. B. 2000b. A model for the interpretation of grooming patterns applied to the Belizean black howling monkey (*Alouatta pigra*). *Primate Rep.* **56**:23–32.

Brockett, R. C., Horwich, R. H., and Jones, C. B. 2000c. Reproductive seasonality in the Belizean black howling monkey (*Alouatta pigra*). *Neotrop. Primates* **8**:136–138.

Brockmann, H. J. 2001. The evolution of alternative strategies and tactics. *Adv. Stud. Behav.* **30**:1–51.

Bronikowski, A. M., and Altmann, J. 1996. Foraging in a variable environment: weather patterns and the behavioral ecology of baboons. *Behav. Ecol. Sociobiol.* **39**:11–25.

Bronstein, J. L. 2001. The costs of mutualism. *Am. Zool.* **41**:825–839.

Bronstein, J. L. 2003. The scope for exploitation within mutualistic interactions. In: P. Hammerstein (ed.), *Genetic and Cultural Evolution of Cooperation*, pp. 185–202, The MIT Press, Cambridge.

Brooks, D. R., and McLennan, D. A. 2002. *The Nature of Diversity: An Evolutionary Voyage of Discovery.* The University of Chicago Press, Chicago.

Brosnan, S. F., and de Waal, F. B. M. 2003. Monkeys reject unequal pay. *Nature* **425**:297–299.

Brown, J. L. 1978. Avian communal breeding systems. *Ann. Rev. Ecol. Syst.* **9**:99–122.

Brown, W. D., Crespi, B. J., and Choe, J. C. 1997. Sexual conflict and the evolution of mating systems. In: J. C. Choe, and B. J. Crespi (eds.), *The Evolution of Mating Systems in Insects and Arachnids*, pp. 352–377, Cambridge University Press, Cambridge.

Buchan, J. C., Alberts, S. C., Silk, and Altmann, J. 2003. True paternal care in a multimale primate society. *Nature* **425**:179–181.

Buss, D. M. 1984. Toward a psychology of person-environment (P-E) correlation: the role of spouse selection. *J. Person. Soc. Psychol.* **47**:361–377.

Byrne, R. W., and Whiten, A. 1988. *Machiavellian Intelligence: Social Expertise and the Evolution of Intellect in Monkeys, Apes and Humans.* Clarendon Press, Oxford.

Byrne, R. W., and Whiten, A. 1997. Machiavellian intelligence. In: A. Whiten, and R. W. Byrne (eds.), *Machiavellian Intelligence II: Extensions and Evaluations*, pp. 1–23, Cambridge University Press, Cambridge.

Call, J., and Tomasello, M. 2003. Social cognition. In: D. Maestripieri (ed.), *Primate Psychology*, pp. 234–253, Harvard University Press, Cambridge.

Carlsbeek, R., Alonzo, S. H., Zamudio, K., and Sinervo, B. 2002. Sexual selection and alternative mating behaviours generate demographic stochasticity in small populations. *Proc. R. Soc. Lond. B* **269**:157–164.

Carpenter, C. R. 1934. A field study of the behavior and social relations of howling monkeys (*Alouatta palliata*). *Comp. Psychol. Monogr.* **10**:1–168.

Carpenter, C. R. 1965. The howlers of Barro Colorado Island. In: I. DeVore (ed.), *Primate Behavior: Field Studies of Monkeys and Apes*, pp. 250–291, Holt, Rinehart, and Winston, New York.

Case, T. J. 2000. *An Illustrated Guide to Theoretical Ecology.* Oxford University Press, New York.

Caughley, G. J. 1964. Density and dispersion of two species of kangaroo in relation to habitat. *Aust. J. Zool.* **12**:238–249.

Chapais, B., Gauthier, C., and Prud'homme, J. 1995. Dominance competition through affiliation and support in Japanese macaques: An experimental study. *Int. J. Primatol.* **16**:521–536.

Chapman, C. A., and Balcomb, S. R. 1998. Population characteristics of howlers: ecological conditions or group history. *Int. J. Primatol.* **19**:385–404.

Chapman, T., Arnqvist, G., Bangham, J., and Rowe, L. 2003a. Sexual conflict. *Trends Ecol. Evol.* **18**:41–47.

Chapman, T., Arnqvist, G., Bangham, J., and Rowe, L. 2003b. Response to Eberhard and Cordero, and Córdoba-Aguilar and Contreras-Garduño: sexual conflict and female choice. *Trends Ecol. Evol.* **18**:440–441.

Chapman, T. W., Crespi, B. J., Kranz, B. D., and Schwarz, M. P. 2000. High relatedness and inbreeding at the origin of eusociality in gall-inducing thrips. *Proc. Nat. Acad. Sci. USA* **97**:1648–1650.

Charnov, E. L. 2002. Reproductive effort, offspring size and benefit-cost ratios in the classification of life histories. *Evol. Ecol. Res.* **4**:749–758.

Chasteen, S. 2003. Sex and gender scientists explore a revolution in evolution. http://www.eurekalert.org/pub_releases/2003-02/su-sag021003.php.

Cheney, D. L., and Seyfarth, R. M. 1982. How vervet monkeys perceive their grunts: field playback experiments. *Anim. Behav.* **30**:739–751.

Chittka, L., Dyer, A. G., Bock, F., and Dornhaus, A. 2003. Bees trade off foraging speed for accuracy. *Nature* **424**:388.

Chivers, D. J. 1969. On the daily behavior and spacing of howling monkey groups. *Folia Primatol.* **10**:48–102.

Chivers, D. J. 1991. Species differences in tolerance to environmental change. In: H. O. Box (ed.), *Primate Responses to Environmental Change*, pp. 5–38, Chapman and Hall, London.

Choe, J. C., and Crespi, B. J. 1997. *The Evolution of Social Behavior in Insects and Arachnids.* Cambridge University Press, Cambridge.

Cialdini, R. B. 2000. *Influence, Science, and Practice*, 4th edn. Allyn & Bacon, Boston.

Cichon, M. 1996. The evolution of brood parasitism: the role of facultative parasitism. *Behav. Ecol.* **7**:137–139.

Clark, A. B. 1991. Individual variation in responsiveness to environmental change. In: H. O. Box (ed.), *Primate Responses to Environmental Change*, pp. 91–110, Chapman and Hall, London.

Clarke, M. R. 1990. Behavioral development and socialization of infants in a free-ranging group of howling monkeys (*Alouatta palliata*). *Folia Primatol.* **54**:1–15.

Clarke, M. R., and Glander, K. E. 1984. Female reproductive success in a group of free-ranging howling monkeys (*Alouatta palliata*) in Costa Rica. In: M. F. Small (ed.), *Female Primates: Studies by Women Primatologists*, pp. 111–126, Alan R. Liss, New York.

Clarke, M. R., and Glander, K. E. 2004. Adult migration patterns of the mantled howlers of La Pacífica. *Am. J. Primatol.* **62**:87 (Abstract).

Clarke, M. R., Collins D. A., and Zucker E. L. 2002. Responses to deforestation in a group of mantled howlers (*Alouatta palliata*) in Costa Rica. *Int. J. Primatol.* **23**:365–381.

Clobert, J., Danchin E., Dhondt A. A., and Nichols, J. D. (eds.) 2001. *Dispersal*. Oxford: Oxford University Press.

Clutton-Brock, T. H. 1974. Primate social organisation and ecology. *Nature* **250**:539–542.

Clutton-Brock, T. H. 1977. Primate ecology and social organisation. *J. Zool.* **183**:1–39.

Clutton-Brock, T. H. 1998. Reproductive concessions and skew in vertebrates. *Trends Ecol. Evol.* **13**:188–292.

Clutton-Brock, T. H. 2002. Breeding together: kin selection and mutualism in cooperative vertebrates. *Science* **296**:69–72.

Clutton-Brock, T. H., Russell, A. F., and Sharpe, L. L. 2003. Meerkat helpers do not specialize in particular activities. *Anim. Behav.* **66**:531–540.

Cole, B. J., and Wiernasz, D. C. 1999. The selective advantage of low relatedness. *Science* **285**:891–893.

Colillas, O., and Coppo, J. 1978. Breeding *Alouatta caraya* in Centro Argentino de Primates. In: D. J. Chivers, and W. Lane-Petter (eds.), *Recent Advances in Primatology*, Vol 2: *Conservation*, pp. 201–214, New York, Academic Press.

Combes, C. 1995. *Interactions Durables: Ecologie et Evolution du Parasitisme*. Editions Masson, Paris.

Connor, R. C., Heithaus, M. R., and Barre, L. M. 1999. Superalliance of bottlenose dolphins. *Nature* **397**:571–572.

Connor, R. C., Heithaus, M. R., and Barre, L. M. 2001. Complex social structure, alliance stability and mating access in a bottlenose dolphin "super-alliance." *Proc. R. Soc. Lond. B* **268**:263–267.

Costa, J. T., and Fitzgerald, T. D. 1996. Developments in social terminology: Semantic battles in a conceptual war. *Trends Ecol. Evol.* **11**:285–289.

Cowlishaw, G., and Dunbar, R. 2000. *Primate Conservation Biology*. The University of Chicago Press, Chicago.

Crabbe, J. C., and Phillips, T. J. 2003. Mother nature meets mother nurture. *Nat. Neurosci.* **5**:440–442.

Craig, J. V., and Guhl, A. M. 1969. Territorial behavior and social interactions of pullets kept in large flocks. *Poultry Sci.* **48**:1622–1628.

Crespi, B. J. 2000. The evolution of maladaptation. *Heredity* **84**:623–629.

Crespi, B. J., and Choe, J. C. 1997a. Explanation and evolution of social systems. In: J. C. Choe, and B. J. Crespi (eds.), *The Evolution of Social Behavior in Insects and Arachnids*, pp. 499–524, Cambridge University Press, New York.

Crespi, B. J., and Choe, J. C. 1997b. Introduction. In: Choe, J. C., and Crespi, B. J. (eds.), *The Evolution of Social Behavior in Insects and Arachnids*, pp. 1–7, Cambridge University Press, New York.

Crespi, B. J., and Yanega, D. 1995. The definition of eusociality. *Behav. Ecol.* **6**:109–115.

Crockett, C. M. 1984. Emigration by female red howler monkeys and the case for female competition. In: M. F. Small (ed.), *Female Primates: Studies by Women Primatologists*, pp. 159–174, Alan R. Liss, New York.

Crockett, C. M. 1998. Conservation biology of the genus *Alouatta*. *Int. J. Primatol.* **19**:549–578.

Crockett, C. M. 2003. Re-evaluating the sexual selection hypothesis for infanticide by *Alouatta* males. In: C. B. Jones (ed.), *Sexual Selection and Reproductive*

Competition in Primates: New Perspectives and Directions, pp. 327–365, American Society of Primatologists, Norman.

Crockett, C. M., and Eisenberg, J. F. 1987. Howlers: Variations in group size and demography. In: B. B. Smuts, D. L. Cheney, R. M. Seyfarth, R. W. Wrangham, and T. T. Struhsaker (eds.), *Primate Societies*, pp. 54–68, The University of Chicago Press, Chicago.

Crockett, C. M., and Janson, C. H. 2000. Infanticide in red howlers: female group size, male membership, and a possible link to folivory. In: C. P. van Schaik, and C. H. Janson (eds.), *Infanticide by Males and its Implications*, pp. 75–98, Cambridge University Press, Cambridge.

Crockett, C. M., and Rudran, R. 1987. Red howler monkey birth data I: seasonal variation. *Am. J. Primatol.* **13**:347–368.

Crook, J. H. 1964. The evolution of social organization and visual communication in the weaver birds (Ploceinae). *Behaviour* **10**:1–178.

Crook, J. H. 1970. Introduction—social behaviour and ethology. In: J. H. Crook (ed.), *Social Behaviour in Birds and Mammals: Essays on the Social Ethology of Animals and Man*, pp. xxi–x1, Academic Press, London.

Crook, J. H. 1971. Sources of cooperation in animals and man. In: J. F. Eisenberg, and W. S. Dillon (eds.), *Man and Beast: Comparative Social Behavior*, pp. 236–260. Smithsonian Institution Press, Washington.

Crook, J. H. 1972. Sexual selection, dimorphism, and social organization in the primates. In: B. Campbell (ed.), *Sexual Selection and the Descent of Man, 1871–1971*, pp. 231–281. Aldine Publishing Co., Chicago

Crook, J. H., and Gartlan, J. S. 1966. Evolution of primate societies. *Nature* **210**:1200–1203.

Crowley, P. H. 2003. Origins of behavioral variability: categorical and discriminative assessment in serial contests. *Anim. Behav.* **66**:427–440.

Daly, M., and Wilson, M. 1988. *Homicide*. Aldine de Gruyter, New York.

Darwin, C. 1859. *The Origin of Species*. The Modern Library, New York.

Darwin, C. 1871. *The Descent of Man*. The Modern Library, New York.

Darwin, C. 1965. *The Expression of the Emotions in Man and Animals*. The University of Chicago Press, Chicago.

Dausmann, K. H., Glos, J., Ganzhorn, J. U., and Heldmaier, G. 2004. Hibernation in a tropical primate. *Nature* **429**:825–826.

Dawkins, R. 1999. *The Extended Phenotype: The Long Reach of the Gene*. Oxford University Press, Oxford.

Dawkins, R., and Krebs, J. R. 1978. Animal signals: Information or manipulation? In: J. R. Krebs, and N. B. Davies (eds.), *Behavioural Ecology: An Evolutionary Approach*, pp. 282–309, Blackwell Scientific Publications, Oxford.

Day, R. L., Coe, R. L., Kendal, J. R., and Laland, K. N. 2003. Neophilia, innovation, and social learning: A study of intergeneric differences in callitrichid monkeys. *Anim. Behav.* **65**:559–571.

Day, T. 2003. Virulence evolution and the timing of disease life-history events. *Trends Ecol. Evol.* **18**:113–118.

De Jaegher, K. 2003. Error-proneness as a handicap signal. *J. Theor. Biol.* **224**:139–152.

de Waal, F. 1989. *Chimpanzee Politics: Power and Sex Among Apes*. The Johns Hopkins University Press, Baltimore.

de Waal, F. 1990. *Peacemaking Among Primates.* Harvard University Press, Cambridge.

de Waal, F. B. M. 1987. Dynamics of social relationships. In: B. B. Smuts, D. L. Cheney, R. M. Seyfarth, R. W. Wrangham, and T. T. Struhsaker (eds.), *Primate Societies,* pp. 421–429. The University of Chicago Press, Chicago.

de Waal, F. B. M. 2000. The first kiss: foundations of conflict resolution research in animals. In: F. Aureli, and F. B. M. de Waal (eds.), *Natural Conflict Resolution,* pp. 15–33. University of California Press, Berkeley.

Deag, J. M., and Crook, J. H. 1971. Social behaviour and "agonistic buffering" in the wild Barbary macaque *Macaca sylvanus* L. *Folia Primatol.* **15**:183–200.

Debat, V., and David, P. 2001. Mapping phenotypes: canalization, plasticity, and developmental stability. *Trends Ecol. Evol.* **16**:555–561.

Deecke, V. B., Slater, P. J. B., and Ford, J. K. B. 2002. Selective habituation shapes acoustic predator recognition in harbour seals. *Nature* **420**:171–173.

DeFries, J. C., Vogler, G. P., and LaBuda, M. C. 1986. Colorado Family Reading Study: an overview. In: J. L. Fuller, and E. C. Simmel (eds.), *Perspectives in Behavior Genetics* pp. 29–56. Erlbaum, Hillsdale.

Dickinson, J. L., and Koenig, W. D. 2003. Desperately seeking similarity. *Science* **300**:1887–1889.

Digby, L. 2000. Infanticide by female mammals: Implications for the evolution of social systems. In: C. P. van Schaik, and C. H. Janson (eds.), *Infanticide by Males and its Implications,* pp. 423–446. Cambridge University Press, Cambridge.

Dillon, N. 2003. Positions, please . . . . *Nature* **425**:457.

Dixson, A. F. 1998. *Primate Sexuality: Comparative Studies of the Prosimians, Monkeys, Apes, and Human Beings.* Cambridge University Press, Cambridge.

Doebeli, M., Blarer, A., and Ackermann, M. 1997. Population dynamics, demographic stochasticity, and the evolution of cooperation. *Proc. Nat. Acad. Sci. USA* **94**:5167–5171.

Double, M. C., and Cockburn, A. 2003. Subordinate superb fairy-wrens (*Malurus cyaneus*) parasitize the reproductive success of attractive dominant males. *Proc. R. Soc. Lond. B* **270**:379–384.

Drickamer, L. C., Gowaty, P. A., and Holmes, C. M. 2000. Free female mate choice in house mice affects reproductive success and offspring viability and performance. *Anim. Behav.* **59**:371–378.

Dugatkin, L. A. 1997. Can kinship ever inhibit the evolution of cooperation? *EcoScience* **4**:460–464.

Dulac, D. and Axel, R. 1995. A novel family of genes encoding putative pheromone receptors in mammals. *Cell* **83**:185–206.

Dumont, Y., Cadieux, A., Doods, H., Pheng, L. H., Abounader, R., Hamel, E., Jacques, D., Regoli, D., and Quirion, R. 2000. BIIE0246, a potent and highly selective non-peptide neuropeptide Y Y(2) receptor antagonist. *Br. J. Pharmacol.* **129**:1075–1088.

Dunbar, R. I. M. 1982. Intraspecific variations in mating strategy. In: Bateson, P. P. G., and Klopfer, P. (eds.), *Perspectives in Ethology,* pp. 385–396, Plenum Press, New York.

Dunbar, R. I. M. 1988. *Primate Social Systems.* Cornell University Press, Ithaca.

Dunbar, R. I. M. 1992. Time: A hidden constraint on the behavioral ecology of baboons. *Behav. Ecol. Sociobiol.* **31**:35–49.

Dunbar, R. I. M. 1996. Determinants of group size in primates: A general model. In: Runciman, W. G., Maynard Smith, J., and Dunbar, R. I. M. (eds.), *Evolution of Social Behaviour Patterns in Primates and Man*, pp. 33–58, Oxford University Press, Oxford.

Dunbar, R. I. M. 1997. *Grooming, Gossip, and the Evolution of Language*. Harvard University Press, Cambridge.

Dunbar, R. I. M. 1998. The social brain hypothesis. *Evol. Anthropol.* **6**:178–190.

Dunbar, R. I. M. 2002. Modelling primate behavioral ecology. *Int. J. Primatol.* **23**:785–819.

Dunbar, R. I. M. 2003. Evolution of the social brain. *Science* **302**:1160–1161.

East, M. L., Burke, T., Wilhelm, K., Greig, C., and Hofer, H. 2003. Sexual conflicts in spotted hyenas: male and female mating tactics and their reproductive outcome with respect to age, social status and tenure. *Proc. R. Soc. Lond. B* **270**:1247–1254.

Eibl-Eibesfeldt, I. 1970. *Ethology: The Biology of Behavior*. Holt, Rinehart and Winston, New York.

Eisenberg, J. F. 1966. The social organizations of mammals. *Handbook Zool.* **10**:1–100.

Eisenberg, J. F. 1979. Habitat, economy, and society: some correlations and hypotheses for the neotropical primates. In: I. S. Bernstein, and E. O. Smith (eds.), *Primate Ecology and Human Origins: Ecological Influences on Social Organization*, pp. 215–262, Garland Press, New York.

Eisenberg, J. F. 1981. *The Mammalian Radiations: An Analysis of Trends in Evolution, Adaptation, and Behavior*. The University of Chicago Press, Chicago.

Eisenberg, J. F., Muckenhirn, N. A., and Rudran, R. 1972. The relation between ecology and social structure of the primates. *Science* **176**:863–874.

Ellefson, J. O. 1968. Territorial behavior in the common white-handed gibbon, *Hylobates lar* Linn. In: P. C. Jay (ed.), *Primates: Studies in Adaptation and Variability*, pp. 180–199, Holt, Rinehart, and Winston, New York.

Emlen, J. M. 1973. *Ecology: An Evolutionary Approach*. Addison-Wesley Publishing Company, Reading.

Emlen, S. T. 1991. Evolution of cooperative breeding in birds and mammals. In: J. R. Krebs, and N. B. Davies (eds.), *Behavioural Ecology: An Evolutionary Approach*, pp. 301–337, Blackwell Scientific Publications, London.

Emlen, S. T. 1994. Benefits, constraints and the evolution of the family. *Trends Ecol. Evol.* **9**:282–284.

Emlen, S. T. 1995. An evolutionary theory of the family. *Proc. Nat. Acad. Scis. USA* **92**:8092–8099.

Emlen, S. T., and Oring, L. W. 1977. Ecology, sexual selection and the evolution of mating systems. *Science* **197**:215–223.

Emlen, S. T., and Wrege, P. H. 1994. Gender, status and family fortunes in the white-fronted bee-eater. *Nature* **367**:129–132.

Enquist, M., and Leimar, O. 1983. Evolution of fighting behaviour: decision rules and assessment of relative strength. *J. Theor. Biol.* **102**:387–410.

Essock-Vitale, S., and Seyfarth, R. M. 1987. Intelligence and social cognition. In: B. B. Smuts, D. L. Cheney, R. M. Seyfarth, R. W. Wrangham, and T. T. Struhsaker (eds.), *Primate Societies*, pp. 452–461, The University of Chicago Press, Chicago.

Estrada, A., Mendoza, A., Castellanos, L., Pacheco, R., Van Belle, S., Garcia, Y., and Munoz, D. 2002. Population of the black howler monkey (*Alouatta pigra*) in a fragmented landscape in Palenque, Chiapas, Mexico. *Am. J. Primatol.* **58**:45–55.

Fagen, R. M. 1980. When doves conspire: evolution of non-damaging fighting tactics in a nonrandom-encounter animal conflict model. *Am. Nat.* **115**:858–869.

Fedigan, L. M., Rose, L. M., and Avila, R. M. 1998. Growth of mantled howler groups in a regenerating Costa Rican dry forest. *Int. J. Primatol.* **19**:405–432.

Ferriere, R., and Le Galliard, J.-F. 2001. Invasion fitness and adaptive dynamics in spatial population models. In: J. Clobert, E. Danchin, A. A. Dhondt, and J. D. Nichols (eds.), *Dispersal*, pp. 57–79, Oxford University Press, Oxford.

Fewell, J. H. 2003. Social insect networks. *Science* **301**:1867–1870.

Fleagle, J. G. 1999. *Primate Adaptation and Evolution*, 2nd edn. Academic Press, San Diego.

Fleagle, J. G., Janson, C., and Reed, K. E. (eds.) 1999. *Primate Communities*. Cambridge University Press, Cambridge.

Fleming, T. H., Breitwisch, R., and Whitesides, G. H. 1987. Patterns of tropical vertebrate frugivore diversity. *Ann. Rev. Ecol. Syst.* **18**:91–109.

Foerster, K., Delhey, K., Johnsen, A., Lifjeld, J. T., and Kempenaers, B. 2003. Females increase offspring heterozygosity and fitness through extra-pair matings. *Nature* **425**:714–717.

Forsgren, E. 1997. Female sand gobies prefer good fathers over dominant males. *Proc. R. Soc. Lond. B* **264**:1283–1286.

Foster, K. R., Wenseleers, T., and Ratnieks, F. L. W. 2001. Spite: Hamilton's unproven theory. *Ann. Zool. Fennici.* **38**:229–238.

Fox, E. A. 2001. Homosexual behavior in wild Sumatran orangutans (*Pongo pygmaeus abelii*). *Am. J. Primatol.* **55**:177–181.

Fragaszy, D. M., and Perry, S. (eds.) 2003a. *The Biology of Traditions: Models and Evidence*. Cambridge University Press, Cambridge.

Fragaszy, D. M., and Perry, S. 2003b. Towards a biology of traditions. In: D. M. Fragaszy, and S. Perry (eds.), *The Biology of Traditions: Models and Evidence*, pp. 1–32, Cambridge University Press, Cambridge.

Fragaszy, D. M., Visalberghi, E., and Robinson, J. G. 1990. Variability and adaptability in the genus Cebus. *Folia Primatol.* **54**:114–118.

Francis, D. D., Szegda, K. Campbell, G., Martin, W. D., and Insel, T. R. 2003. Epigenetic sources of behavioral differences in mice. *Nat. Neurosci.* **6**:445–446.

Frank, S. A. 1998. *Foundations of Social Evolution*. Princeton University Press, Princeton.

Frank, S. A. 1995. Mutual policing and repression of competition in the evolution of cooperative groups. *Nature* **377**:520–522.

Frankie, G. W., Baker, H. G., and Opler, P. A. 1974. Comparative phenological studies of trees in tropical wet and dry forests in the lowlands of Costa Rica. *J. Ecol.* **62**:881–919.

French, J. A. 1997. Proximate regulation of singular breeding in callitrichid primates. In: N. G. Solomon, and J. A. French (eds.), *Cooperative Breeding in Mammals*, pp. 34–75, Cambridge University Press, Cambridge.

Frith, C. D., and Frith, U. 1999. Interacting minds—a biological basis. *Science* **286**:1692–1695.

Fukuda, F. 1983. Troop desertion by female Japanese macaques. *Jap. J. Ecol.* **33**:347–355 (in Japanese with English abstract).

Fukuda, F. 1988. Influence of artificial food supply on population parameters and dispersal in the Hakone T Troop of Japanese macaques. *Primates* **29**:477–492.

Fukuda, F. 2004. Dispersal and environmental disturbance in Japanese macaques (*Macaca fuscata*). *Primate Rep.* **68**:53–69.

Gadagkar, R. 1994. Why the definition of eusociality is not helpful to understand its evolution and what should we do about it. *Oikos* **70**:485–487.

Gadagkar, R. 1997. *Survival Strategies: Cooperation and Conflict in Animal Societies.* Harvard University Press, Cambridge.

Gagneux, P. 2002. The genus *Pan*: Population genetics of an endangered outgroup. *Trends Genet.* **18**:327–330.

Gaines, M. S., and Whittam, T. S. 1980. Genetic changes in fluctuating vole populations: selective vs. nonselective forces. *Genetics* **96**:767–778.

Galef, B. G. 1981. The ecology of weaning: Parasitism and the achievement of independence by altricial mammals. In: I. Gubernick, and D. J. H. Klopfer (eds.), *Parental Care in Mammals*, pp. 211–241, Plenum, New York.

Gandon, S., and Michalakis Y. 2001. Multiple causes of the evolution of dispersal. In: J. Clobert, E. Danchin, A. A. Dhondt, and J. D. Nichols (eds.), *Dispersal*, pp. 155–167, Oxford University Press, Oxford.

Garber, P. A., and Leigh, S. R. 1997. Ontogenetic variation in small-bodied New World primates: Implications for patterns of reproduction and infant care. *Folia Primatol.* **68**:1–22.

Gavrilets, S. 2000. Rapid evolution of reproductive barriers driven by sexual conflict. *Nature* **403**:886–889.

Gavrilets, S., and Waxman, D. 2002. Sympatric speciation by sexual conflict. *Proc. Nat. Acad. Scis. USA* **99**:10533–10538.

Gavrilets, S., Arnqvist, G., and Friberg, U. 2001. The evolution of female mate choice by sexual conflict. *Proc. R. Soc. Lond. B* **268**:531–539.

Gehring, W. J., and Willoughby, A. R. 2002. The medial frontal cortex and the rapid processing of monetary gains and losses. *Science* **295**:2279–2282.

Geisel, J. T. 1976. Reproductive strategies as adaptations to life in temporally heterogeneous environments. *Ann. Rev. Ecol. Syst.* **7**:7–80.

Gerald, M. S. 2003. How color may guide the primate world: possible relationships between sexual selection and sexual dichromatism. In: C. B. Jones (ed.), *Sexual Selection and Reproductive Competition in Primates: New Perspectives and Directions*, pp. 141–171, American Society of Primatologists, Norman.

German, R. Z., Hertweck, D. W., Sirianni, J. E., and Swindler, D. R. 1994. Heterochrony and sexual dimorphism in the pigtailed macaque (*Macaca nemestrina*). *Am. J. Phys. Anthropol.* **93**:373–380.

Gewin, V. 2003. A plea for diversity. *Nature* **422**:368–369.

Ghalambor, C. K., and Martin, T. E. 2002. Comparative manipulation of predation risk in incubating birds reveals variability in the plasticity of responses. *Behav. Ecol.* **13**:101–108.

Gillespie, J. 1974. The role of environmental grain in the maintenance of genetic variation. *Am. Nat.* **108**:831–836.

Ginther, A. J., Ziegler, T. E., and Snowdon, C. T. 2001. Reproductive biology of captive male cottontop tamarin monkeys as a function of social environment. *Anim. Behav.* **61**:65–78.

Giraud, T., Pedersen, J. S., and Keller, L. 2002. Evolution of supercolonies: the Argentine ants of southern Europe. *Proc. Nat. Acad. Scis. USA* **99**:6075–6079.

Glander, K. E. 1975. *Habitat and Resource Utilization: An Ecological View of Social Organization in Mantled Howling Monkeys.* Ph.D. dissertation, University of Chicago (Unpublished).

Glander, K. E. 1980. Reproduction and population growth in free-ranging mantled howling monkeys. *Am. J. Phys. Anthropol.* **53**:25–36.

Glander, K. E. 1992. Dispersal patterns in Costa Rican mantled howling monkeys. *Int. J. Primatol.* **13**:415–436.

Glimcher, P. W. 2003. *Decisions, Uncertainty, and the Brain: The Science of Neuroeconomics.* MIT Press, Cambridge.

Godfray, H. J. C., and Johnstone, R. A. 2000. Begging and bleating: the evolution of parent-offspring signalling. *Philos. Trans. R. Soc. B* **355**:1581–1591.

Godfrey, L. R., and Sutherland, M. R. 1996. Paradox of peramorphic paedomorphosis: Heterochrony and human evolution. *Am. J. Phys. Anthropol.* **99**:17–42.

Goldizen, A. W. 1987. Tamarins and marmosets: communal care of offspring. In: B. B. Smuts, D. L. Cheney, R. M. Seyfarth, R. W. Wrangham, and T. T. Struhsaker (eds.), *Primate Societies*, pp. 34–43, The University of Chicago Press, Chicago.

Goldsmith, H., Gernsbacher, M. A., Crabbe, J., Dawson, G., Gottesman, I. I., Hewitt, J., McGue, M., Pedersen, N., Plomin, R., Rose, R., and Swanson, J. 2003. Research psychologists' roles in the genetic revolution. *Am. Psychol.* **58**:318–319.

Goodhill, G. J., Bates, K. R., and Montague, P. R. 1997. Influences on the global structure of cortical maps. *Proc. R. Soc. Lond. B* **22**:649–655.

Gosling, S. D., Lilienfeld, S. O., and Marino, L. 2003. Personality. In: D. Maestripieri (ed.), *Primate Psychology*, pp. 254–288, Harvard University Press, Cambridge.

Gowaty, P. A. 1997. Sexual dialectics, sexual selection, and variation in reproductive behavior. In: P. A. Gowaty (ed.), *Feminism and Evolutionary Biology: Boundaries, Intersections, and Frontiers*, pp. 351–384, Chapman and Hall, New York.

Goymann, W., Eat, M. L., and Hofer, H. 2001. Androgens and the role of female "hyperaggressiveness" in spotted hyenas (*Crocuta crocuta*). *Horm. Behav.* **39**:83–92.

Grafen, A. 2002. A state-free optimization model for sequences of behavior. *Anim. Behav.* **63**:183–191.

Grammer, K. 1989. Human courtship behaviour: biological basis and cognitive processing. In: A. E. Rasa, C. Vogel, and E. Voland (eds.), *The Sociobiology of Sexual and Reproductive Strategies*, pp. 147–169, Chapman and Hall, New York.

Grieco, F., van Noordwijk, A. J., and Visser, M. E. 2002. Evidence for the effect of learning on timing of reproduction in blue tits. *Science* **296**:136–138.

Griffin, A. S., and West, S. A. 2003. Kin discrimination and the benefit of helping in cooperatively breeding vertebrates. *Science* **302**:634–636.

Gross, M. R. 1996. Alternative reproductive strategies and tactics: diversity within sexes. *Trends Ecol. Evol.* **11**:92–98.

Groves, C. 2001. *Primate Taxonomy.* Smithsonian Institution Press, Washington.

Haccou, P., Glaizot, O., and Cannings, C. 2003. Patch leaving strategies and superparatism: an asymmetric generalized war of attrition. *J. Theor. Biol.* **225**:77–89.

Hager, R. 2003a. Models of reproductive skew applied to primates. In: C. B. Jones (ed.), *Sexual Selection and Reproductive Competition in Primates: New Perspectives and Directions*, pp. 65–101, American Society of Primatologists, Norman.

Hager, R. 2003b. The effects of dispersal costs on reproductive skew and within-group aggression in primate groups. *Primate Rep.* **67**:85–98.

Hager, R. 2003c. Behavioural ecology. *Encyclopedia of Life Sciences.* Nature Publishing Group, London.

Hager, R., and Johnstone, R. 2003. The genetic basis of family conflict resolution in mice. *Nature* **421**:533–535.

Hager, R., and Welker, C. 2001. Female dominance in African lorises (*Otolemur garnettii*). *Folia Primatol.* **72**:48–50.

Haig, D., and Graham, C. 1991. Genomic imprinting and the strange case of the insulin-like growth factor II receptor. *Cell* **64**:1045–1046.

Halushka, M. K., Fan, J. B., Bentley, K., Hsie, L., Shen, N., Weder, A., Cooper, R., Lipshutz, R., and Chakravarti, A. 1999. Patterns of single-nucleotide polymorphisms in candidate genes for blood-pressure homeostasis. *Nat. Genet.* **22**:239–247.

Hamilton, I. M., and Dill, L. M. 2002. Three-player social parasitism games: implications for resource defense and group formation. *Am. Nat.* **159**:670–686.

Hamilton, W. D. 1964. The evolution of social behavior. *J. Theor. Biol.* **7**:1–52.

Hamilton, W. D. 1966. The moulding of senescence by natural selection. *J. Theor. Biol.* **12**:12–45.

Hanski, I. 1994. Patch-occupancy dynamics in fragmented landscapes. *Trends Ecol. Evol.* **9**:131–135.

Hanski, I. 2001. Population dynamic consequences of dispersal in local populations and in metapopulations. In: J. Clobert, E. Danchin, A. A. Dhondt, and J. D. Nichols (eds.), *Dispersal*, pp. 283–298, Oxford University Press, Oxford.

Hanski, I. A., and Gilpin, M. E. (eds.) 1997. *Metapopulation Biology: Ecology, Genetics, and Evolution.* Academic Press, New York.

Harcourt, A. H. 1998. Sperm competition in primates. *Am. Nat.* **189**:189–194.

Harcourt, A. H., Coppeto S. A., and Parks S. A. 2002. Rarity, specialization and extinction in primates. *J. Biogeog.* **29**:445–456.

Hauber, M. E., and Sherman, P. W. 2001. Self-referent phenotype matching: theoretical considerations and empirical evidence. *Trends Neurosci.* **24**:609–616.

Hausfater, G. 1975. *Dominance and Reproduction in Baboons (Papio cynocephalus).* S. Karger, Basel.

Hausfater, G., Cairns, S. J., and Levin, R. N. 1987. Variability and stability in the rank relations of nonhuman primate females: Analysis by computer simulation. *Am. J. Primatol.* **12**:55–70.

Heckhausen, J., and Singer, T. 2001. Plasticity in human behavior across the lifespan. In: N. J. Smelser, and P. B. Balter (eds.), *International Encyclopedia of the Social and Behavioral Sciences*, pp. 11497–11501, Elsevier Science Ltd., New York.

Heinze, J., and Keller, L. 2000. Alternative reproductive strategies: a queen perspective in ants. *Trends Ecol. Evol.* **15**:508–512.

Helms Cahan, S., Blumstein, D. T., Sundström, L., Liebig, J., and Griffin, A. 2002. Social trajectories and the evolution of social behavior. *Oikos* **96**:206–216.

Helms Cahan, S., and Keller, L. 2003. Complex hybrid origin of genetic caste determination in harvester ants. *Nature* **424**:306–309.

Henikoff, S. 2003. Versatile assembler. *Nature* **423**:814–817.

Herrnstein, R. J., and Loveland, D. H. 1964. Complex visual concept in the pigeon. *Science* **146**:549–551.

Hewitt, G. M., and Butlin, R. K. 1997. Causes and consequences of population structure. In: J. R. Krebs, and N. B. Davies (eds.), *Behavioural Ecology: An Evolutionary Approach*, pp. 350–372, Oxford University Press, Oxford.

Hewitt, J. K., and Turner, J. R. 1995. Behavior genetic studies of cardiovascular responses to stress. In: J. R. Turner, L. R. Cardon, and J. K. Hewitt (eds.), *Behavior Genetic Approaches in Behavioral Medicine*, pp. 87–103, Plenum, New York.

Hochachka, P. W., and Somero, G. N. 1973. *Strategies of Biochemical Adaptation*. W. B. Saunders Co., Philadelphia.

Hoffmann, A. A., Hallas, R. J., Dean, J. A., and Schiffer, M. 2003. Low potential for climatic stress adaptation in a rainforest *Drosophila* species. *Science* **301**:100–103.

Holland, B., and Rice, W. R. 1997. Cryptic sexual selection: More control issues. *Evolution* **51**:321–324.

Holland, B., and Rice, W. R. 1999. Experimental removal of sexual selection reverses intersexual antagonistic coevolution and removes a reproductive load. *Proc. Natl. Acad. Sci. USA.* **96**:5083–5088.

Hölldobler, B., and Wilson, E. O. 1990. *The Ants*. The Belknap Press, Cambridge.

Hollis, K. L., Pharr, V. L., Dumas, M. J., Britton, G. B., and Field, J. 1997. Classical conditioning provides paternity advantage for territorial male blue gouramis (*Trichogaster trichopterus*). *J. Comp. Psychol.* **111**:219–225.

Horwich, R. H., Brockett, R. C., and Jones, C. B. 2000. Alternative male reproductive behaviors in the Belizean black howler monkey (*Alouatta pigra*). *Neotrop. Primates* **8**:95–98.

Horwich, R. H., Brockett, R. C., James, R. A., and Jones, C. B. 2001. Population structure and group productivity of the Belizean black howling monkey (*Alouatta pigra*): implications for female socioecology. *Primate Rep.* **61**:47–65.

Hrdy, S. B. 1976. The care and exploitation of nonhuman primate infants by conspecifics other than the mother. *Adv. Stud. Behav.* **6**:101–158.

Hrdy, S. B. 1977. *The Langurs of Abu: Female and Male Strategies of Reproduction*. Harvard University Press, Cambridge.

Hrdy, S. B. 1979. Infanticide among animals: a review, classification, and examination of the implications for the reproductive strategies of females. *Ethol. Sociobiol.* **1**:13–40.

Hrdy, S. B. 1999a. *The Woman That Never Evolved*, 2nd edn. Harvard University Press, Cambridge.

Hrdy, S. B. 1999b. *Mother Nature: Maternal Instincts and How They Shape the Human Species*. Ballantine Books, New York.

Hrdy, S. B., and Hrdy, D. B. 1976. Hierarchical relations among female hanuman langurs (Primates: Colobinae, *Presbytis entellus*. *Science* **197**:913–915.

Hubbell, S. P., Foster, R. B., O'Brien, S., Wechsler, B., Condit, R., Wright, K., and Loo de Lau, S. 1999. Light gaps, recruitment limitation and tree diversity in a Neotropical forest. *Science* **283**:554–557.

Huey, R. B., Hertz, P. E., and Sinervo, B. 2003. Behavioral drive versus behavioral inertia in evolution: a null model approach. *Am. Nat.* **161**:357–366.

Huggins, G. R., and Preti, G. 1981. Vaginal odors and secretions. *Clin. Obstet. Gynecol.* **24**:355–377.

Humle, T., and Matsuzawa, T. 2001. Behavioural diversity among the wild chimpanzee populations of Bossou and neighbouring areas, Guinea and Cote d'Ivoire, West Africa: a preliminary report. *Folia Primatol.* **72**:57–68.

Humle, T., and Matsuzawa, T. 2002. Ant-dipping among the chimpanzees of Bossou, Guinea, and some comparisons with other sites. *Am. J. Primatol.* **58**:133–148.

Hurd, P. L. 1995. Communication in discrete action-response games. *J. Theor. Biol.* **174**:217–222.

Hutchinson, G. E. 1959. Homage to Santa Rosalia or why are there so many kinds of animals? *Am. Nat.* **93**:145–159.

Isles, A. R., and Wilkinson, L. S. 2000. Imprinted genes, cognition, and behavior. *Trends Cogn. Sci.* **4**:309–318.

Jack, K. M. 2003. Explaining variation in affiliative relationships among male white-faced capuchins (*Cebus capucinus*). *Folia Primatol.* **74**:1–16.

Jamieson, I. G., McRae, S. B., Trewby, M., and Simmons, R. E. 2000. High rates of conspecific brood parasitism and egg rejection in coots and moorhens in ephemeral wetlands in Namibia. *Auk* **117**:250–252.

Jeanne, R. L. 1972. Social biology of the Neotropical wasp *Mischocyttarus drewseni*. *Bull. Mus. Comp. Zool. Harv.* **144**:63–150.

Jenkins, E. V., Morris, C., and Blackman, S. 2000. Delayed benefits of paternal care in the burying beetle Nicrophorus vespilloides. *Anim. Behav.* **60**:443–451.

Jennions, M. D., and MacDonald, D. W. 1994. Cooperative breeding in mammals. *Trends Ecol. Evol.* **9**:89–93.

Johnsen, A., Andersen, V., Sunding, C., and Lifjeld, J. T. 2000. Female bluethroats enhance offspring immunocompetence through extra-pair copulations. *Nature* **406**:296–299.

Johnson, D., Macdonald, D., Kays, R., and Blackwell, P. G. 2003. Response to Revilla, and Buckley and Ruxton: The resource dispersion hypothesis. *Trends Ecol. Evol.* **18**:381–382.

Johnson, D. D. P., Kays, R., Blackwell, P. G., and Macdonald, D. W. 2002. Does the resource dispersion hypothesis explain group living? *Trends Ecol. Evol.* **17**:563–570.

Johnson, M. L., and Gaines M. S. 1990. Evolution of dispersal: Theoretical models and empirical tests using birds and mammals. *Ann. Rev. Ecol. Syst.* **21**:449–480.

Johnstone, R. A., and Bshary, R. 2002. From parasitism to mutualism: partner control in asymmetric interactions. *Ecol. Lett.* **5**:634–639.

Johnstone, R. A., and Cant, M. A. 1999. Reproductive skew and indiscriminate infanticide. *Anim. Behav.* **57**:243–249.

Johnstone, R. A., and Earn, D. J. D. 1999. Imperfect female choice and male mating skew on leks of different sizes. *Behav. Ecol. Sociobiol.* **45**:277–281.

Johnstone, R. A., and Keller, L. 2000. How males can gain by harming their mates: sexual conflict, seminal toxins, and the cost of mating. *Am. Nat.* **156**:368–377.

Johnstone, R. A., and Roulin, A. 2003. Sibling negotiation. *Behav. Ecol.* **14**:780–786.

Jones, C. B. 1978. *Aspects of Reproduction in the Mantled Howler Monkey* (*Alouatta palliata* Gray). Ph.D. dissertation, Cornell University (Unpublished).

Jones, C. B. 1979. Grooming in the mantled howler monkey (*Alouatta palliata* Gray). *Primates* **20**:289–292.

Jones, C. B. 1980. The functions of status in the mantled howler monkey (*Alouatta palliata* Gray): intraspecific competition for group membership in a folivorous Neotropical primate. *Primates* **21**:389–405.

Jones, C. B. 1981. The evolution and socioecology of dominance in primate groups: a theoretical formulation, classification, and assessment. *Primates* **22**: 70–83.

Jones, C. B. 1982a. A field manipulation of spatial relations among male mantled howler monkeys. *Primates* **23**:130–134.

Jones, C. B. 1982b. A comment on the selective advantage of male subordination to females in primates ("female dominance"). *Hum. Ethol. Newsl.* **3**:23–26.

Jones, C. B. 1983a. Social organization of captive black howler monkeys (*Alouatta caraya*): "social competition" and the use of non-damaging behavior. *Primates* **24**:25–39.

Jones, C. B. 1983b. Do howler monkeys feed upon legume flowers preferentially at flower opening time? *Brenesia* **21**:41–46.

Jones, C. B. 1983c. Are female cercopithecines altruistic to their daughters? *Hum. Ethol. Newsl.* **4**:17–21.

Jones, C. B. 1985a. Reproductive patterns in mantled howler monkeys: estrus, mate choice, and copulation. *Primates* **26**:130–142.

Jones, C. B. 1985b. "Nice" guys may not finish last?! *Pol. Life Sci.* **4**:82–86 (Book Review).

Jones, C. B. 1986. Infant transfer behavior in humans: a note on the exploitation of young. *Aggress. Behav.* **12**:167–173.

Jones, C. B. 1987. Evidence supporting the Pleistocene forest refuge hypothesis for primates. *Biotropica* **19**:373–375.

Jones, C. B. 1995a. Alternative reproductive behaviors in the mantled howler monkey (*Alouatta palliata* Gray): testing Carpenter's hypothesis. *Bol. Primatol. Lat.* **5**:1–5.

Jones, C. B. 1995b. Howler monkeys appear to be preadapted to cope with habitat fragmentation. *Endangered Species UPDATE* **12**:9–10.

Jones, C. B. 1995c. Dispersal in mantled howler monkeys: a threshold model. *Mastozool. Neotrop.* **2**:207–211.

Jones, C. B. 1995d. The potential for metacommunity effects upon howler monkeys. *Neotrop. Primates* **3**:43–45.

Jones, C. B. 1996a. Temporal division of labor in a primate: age-dependent foraging behavior. *Neotrop. Primates* **4**:50–53.

Jones, C. B. 1996b. Predictability of plant food resources for mantled howler monkeys at Hacienda La Pacifica, Costa Rica: Glander's dissertation revisited. *Neotrop. Primates* **4**:147–149.

Jones, C. B. 1996c. The selective advantage of patriarchal restraint. *Hum. Nat.* **7**:97–102.

Jones, C. B. 1997a. Social parasitism in the mantled howler monkey, *Alouatta palliata* Gray (Primates: Cebidae): involuntary altruism in a mammal? *Sociobiology.* **30**:51–61.

Jones, C. B. 1997b. Life history patterns of howler monkeys in a time-varying environment. *Bol. Primatol. Lat.* **6**:1–8.

Jones, C. B. 1997c. Rarity in primates: implications for conservation. *Mastozool. Neotrop.* **4**:35–47.

Jones, C. B. 1997d. Subspecific differences in vulva size between *Alouatta palliata palliata* and *A. p. mexicana*: implications for assessment of female receptivity. *Neotrop. Primates* **5**:46–48.

Jones, C. B. 1998. A broad-band contact call by female mantled howler monkeys: implications for heterogeneous conditions. *Neotrop. Primates* **6**:38–40.

Jones, C. B. 1999a. Why both sexes leave: effects of habitat fragmentation on dispersal behavior. *Endandered Species UPDATE* **16**:70–73.

Jones, C. B. 1999b. A method to determine when active translocation of nonhuman primates is justified. *J. App. Anim. Wel. Sci.* **2**:229–238.

Jones, C. B. 1999c. Testis symmetry in the mantled howling monkey. *Neotrop. Primates* **7**:117–119.

Jones, C. B. 2000. *Alouatta palliata* politics: empirical and theoretical aspects of power. *Primate Rep.* **56**:3–21.

Jones, C. B. 2001. Introduction: sampling neotropical primates—implications for conservation and socioecology. *Primate Rep.* **61**:3–7.

Jones, C. B. 2002a. Negative reinforcement in primate societies related to aggressive restraint. *Folia Primatol.* **73**:140–143.

Jones, C. B. 2002b. Genital displays by adult male and female mantled howling monkeys, *Alouatta palliata* (Atelidae): evidence for condition-dependent compound displays. *Neotrop. Primates* **10**:144–147.

Jones, C. B. 2002c. A possible example of coercive mating in mantled howling monkeys (*Alouatta palliata*) related to sperm competition. *Neotrop. Primates* **10**:95–96.

Jones, C. B. 2003a. Urine-washing behaviors as condition-dependent signals of quality by adult mantled howler monkeys (*Alouatta palliata*). *Lab. Prim. News.* **42**:12–14.

Jones, C. B. 2003b. *Sexual Selection and Reproductive Competition in Primates: New Perspectives and Directions*. American Society of Primatologists, Norman.

Jones, C. B. 2004. The number of adult females in groups of polygynous howling monkeys (*Alouatta* spp.): theoretical inferences. *Primate Rep.* **68**:7–25.

Jones, C. B., and Agoramoorthy, G. 2003. Alternative reproductive behaviors in primates: towards general principles. In: C. B. Jones (ed.), *Sexual Selection and Reproductive Competition in Primates: New Perspectives and Directions*, pp. 103–139, American Society of Primatologists, Norman.

Jones, C. B., and Cortés-Ortiz, L. 1998. Facultative polyandry in the howling monkey (*Alouatta palliata*): Carpenter was correct. *Bol. Primatol. Lat.* **7**:1–7.

Kameda, T., and Nakanishi, D. 2002. Cost-benefit analysis of social/cultural learning in a nonstationary uncertain environment: an evolutionary simulation and an experiment with human subjects. *Evol. Hum. Behav.* **23**:373–394.

Kamilar, J. 2003. Differential levels of plasticity in cercopithecoid primates. *Am. J. Primatol.* **60** (Suppl. 1):44 (Abstract).

Kanthaswamy, S., and Smith, D. G. 2002. Population subdivision and gene flow among wild orangutans. *Primates* **43**:315–327.

Kappeler, P. M. 1993. Female dominance in primates and other mammals. *Perspect. Ethol.* **10**: 143–158.

Kappeler, P. M. 1999. Primate socioecology: new insights from males. *Naturwissenschaften* **86**:18–29.

Kappeler, P. M., and Pereira, M. E. (eds.). 2003. *Primate Life Histories and Socioecology*. The University of Chicago Press, Chicago.

Kappeler, P. M., Pereira, M. E., and van Schaik, C. P. 2003. Primate life histories and socioecology. In: P. M. Kappeler, and M. E. Pereira (eds.), *Primate Life Histories and Socioecology*, pp. 1–20, The University of Chicago Press, Chicago.

Kawecki, T. J. 1994. Accumulation of deleterious mutations and the evolutionary cost of being a generalist. *Am. Nat.* **144**:833–838.

Kawai, M. 1965. Newly acquired pre-cultural behavior of the natural troop of Japanese monkeys on Koshima Islet. *Primates* **1**:1–30.

Keller, L. 2003. Behavioral plasticity: levels of sociality in bees. *Curr. Biol.* **13**:R644–R645.

Keller, L., and Reeve, H. K. 1994. Partitioning of reproduction in animal societies. *Trends Ecol. Evol.* **9**:98–102.

Kinzey, W. G. 1982. Distribution of primates and forest refuges. In: G. T. Prance (ed.), *Biological Diversification in the Tropics*, pp. 455–482, Columbia University Press, New York.

Kirschner, M. W., and Gerhart, J. 1998. Evolvability. *Proc. Nat. Acad. Scis. USA* **95**:8420–8477.

Kitchen, D. M., and Packer, C. 1999. Complexity in vertebrate societies: In: L. Keller (ed.), *Levels of Selection in Evolution*, pp. 176–196, Princeton University Press, Princeton.

Koenig, A. 2002. Competition for resources and its behavioral consequences among female primates. *Int. J. Primatol.* **23**:759–783.

Kosslyn, S. M., Cacioppo, J. T., Davidson, R. J., Hugdahl, K., Lovallo, W. R., Spiegel, D., and Rose, R. 2002. Bridging psychology and biology: the analysis of individuals in groups. *Am. Psychol.* **57**:341–351.

Kotiaho, J. S., Simmons, L. W., Hunt, J., and Tomkins, J. L. 2003. Males influence maternal effects that promote sexual selection: a quantitative genetic experiment with dung beetles *Onthophagus taurus. Am. Nat.* **161**:852–859.

Kowalewski, M., and Zunino, G. E. 2004. Birth seasonality in *Alouatta caraya* in Northern Argentina. *Int. J. Primatol.* **25**:383–400.

Kowalski, R. M. 2003. *Complaining, Teasing, and Other Annoying Behaviors.* Yale University Press, New Haven.

Kraus, C., Heistermann, M., and Kappeler, P. M. 1999. Physiological suppression of sexual function of subordinate males: a subtle form of intrasexual competition among male sifakas (*Propithecus verreauxi*)? *Physiol. Behav.* **66**:855–861.

Krebs, J. R., and Davies, N. B. 1993. *An Introduction to Behavioural Ecology*, 3rd edn. Blackwell Scientific Publications, Oxford.

Krebs, J. R., and Dawkins, R. 1984. Animal signals: mindreading and manipulation. In: J. R. Krebs, and N. B. Davies (eds.), *Behavioural Ecology: An Evolutionary Approach*, pp. 380–402, 2nd edn. Blackwell Scientific Publications, Oxford.

Krishnamani, R., and Mahaney, W. C. 2000. Geophagy among primates: adaptive significance and ecological consequences. *Anim. Behav.* **59**:899–915.

Kudo, H., and Dunbar, R. I. M. 2001. Neocortex size and social network size in primates. *Anim. Behav.* **62**:711–722.

Kuester, J., and Paul, A. 2000. The use of infants to buffer male aggression. In: F. Aureli, and F. B. M. de Waal, (eds.), *Natural Conflict Resolution*, pp. 91–93, University of California Press, Berkeley.

Kummer, H. 1968. *Social Organization of Hamadryas Baboons.* The University of Chicago Press, Chicago.

Kummer, H. 1995. *In Quest of the Sacred Baboon: A Scientist's Journey.* Princeton University Press, Princeton.

Kummer, H., and Goodall, J. 1985. Conditions of innovative behaviour in primates. *Phil. Trans. R. Soc. Lond. B* **308**:203–214.

Lacey, E. A., and Sherman, P. W. 1997. Cooperative breeding in naked mole rats: Implications for vertebrate and invertebrate sociality. In: N. G. Solomon, and J. A. French (eds.), *Cooperative Breeding in Mammals*, pp. 267–301. Cambridge University Press, Cambridge.

Lande, R. 1980. Genetic variation and phenotypic evolution during allopatric speciation. *Am. Nat.* **122**:114–131.

Lande, R., and Arnold, S. J. 1983. The measurement of selection on correlated characters. *Evolution* **37**:1210–1226.

Langer, P., Hogendoorn, K., and Keller, L. 2004. Tug-of-war over reproduction in a social bee. *Nature* **428**:844–847.

Lee, P. C., and Kappeler, P. M. 2003. Socioecological correlates of phenotypic plasticity of primate life histories. In: P. M. Kappeler, and M. E. Pereira (eds.), *Primate Life Histories and Socioecology*, pp. 41–65, The University of Chicago Press, Chicago.

Lee, R. J. 1997. The Impact of Hunting and Habitat Disturbance on the Population Dynamics and Behavioral Ecology of the Crested Black Macaque (*Macaca nigra*). Ph.D. Dissertation, University of Oregon (Unpublished).

Leimar, O., and Enquist, M. 1984. Effects of asymmetries on owner-intruder conflicts. *J. Theor. Biol.* **111**:475–491.

Lenoir, A., D'Ettorre, P., and Errard, C. 2001. Chemical ecology and social parasitism in ants. *Ann. Rev. Entomol.* **46**:573–599.

Lerner, I. M. 1970. *Genetic Homeostasis.* Dover, New York.

Lessells, C. M. 2002. Parentally biased favouritism: why should parents specialize in caring for different offspring? *Philos. Trans. R. Soc. Lond. B* **357**:381–403.

Levin, S. A. 1976. Population dynamics models in heterogeneous environments. *Ann. Rev. Ecol. Syst.* **7**:287–310.

Levins, R. 1968. *Evolution in Changing Environments.* Princeton University Press, Princeton.

Lewis, S. E., and Pusey, A. E. 1997. Factors influencing the occurrence of communal care in plural bleeding mammals. In: N. G. Solomon, and J. A. French (eds.), *Cooperative Breeding in Mammals*, pp. 335–363, Cambridge University Press, Cambridge.

Lewontin, R. C. 1957. The adaptations of populations to varying environments. *Cold Spring Harbor Symp. Quant. Biol.* **22**:395–408.

Lewontin, R. C. 1974. *The Genetic Basis of Evolutionary Change.* Columbia University Press, New York.

Lewontin, R. C. 2000. *The Triple Helix: Gene, Organism, and Environment.* Harvard University Press, Cambridge.

Li, L. L., Keverne, E. B., Aparicio, S. A., Ishino, F., Barton, S. C., and Surani, M. A. 1999. Regulation of maternal behaviour and offspring growth by paternally expressed gene Peg3. *Science* **284**:330–333.

Lidicker, W. Z., Jr. 1981. Organization and chaos in population structure: Some thoughts on future directions for mammalian population genetics. In: M. H. Smith, and J. Joule (eds.), *Mammalian Population Genetics*, pp. 309–320, University of Georgia, Athens.

Lim, M. M., Wang, Z., Olazábal, D. E., Ren, X., Terwilliger, E. F., and Young, L. J. 2004. Enhanced partner preference in a promiscuous species by manipulating the expression of a single gene. *Nature* **429**:754–758.

Lindenfors, P., Fröberg, L., and Nunn, C. L. 2004. Females drive primate social evolution. *Biol. Lett.* **271**:S101–S103.

Linklater, W. L., and Cameron, E. Z. 2000. Tests for cooperative behavior between stallions. *Anim. Behav.* **60**:731–743.

Little, A. C., Penton-Voak, I. S., Burt, D. M., and Perrett, D. I. 2003. Investigating an imprinting-like phenomenon in humans: Partners and opposite-sex parents have similar hair and eye colour. *Evol. Hum. Behav.* **24**:43–52.

Lorch, P. D., and Chao, L. 2003. Selection for multiple mating in females due to mates that reduce female fitness. *Behav. Ecol.* **14**:679–686.

Luttbeg, B. 2004. Female mate assessment and choice behavior affect the frequency of alternative male mating tactics. *Behav. Ecol.* **15**:239–247.

Mac Arthur, R. H., and Wilson, E. O. 1967. *The Theory of Island Biogeography*. Princeton University Press, Princeton.

Macdonald, D. W., and Johnson, D. D. P. 2001. Dispersal in theory and practice: Consequences for conservation biology. In: J. Clobert, E. Danchin, A. A. Dhondt, and J. D. Nichols. (eds.), *Dispersal*. Oxford University Press, Oxford.

Macy, J. D., Jr., Beattie, T. A., Morgenstern, S. E., and Arnsten, A. F. 2000. Use of guanfacine to control self-injurious behavior in two rhesus macaques (*Macaca mulatta*) and one baboon (*Papio anubis*). *Compar. Med.* **50**:419–425.

Maestripieri, D. 2003a. *Primate Psychology*. Harvard University Press, Cambridge.

Maestripieri, D. 2003b. Similarities in affiliation and aggression between cross-fostered Rhesus macaque females and their biological mothers. *Develop. Psychobiol.* **43**:321–327.

Malmgren, L. A. 1979. *Empirical Population Genetics of Golden Mantled Howling Monkeys (Alouatta palliata) in Relation to Population Structure, Social Dynamics, and Evolution.* Ph.D. dissertation, University of Connecticut (Unpublished).

Maly, I. P., and Sasse, D. 1987. The intra-acinar distribution patterns of alcohol-dehydrogenase activity in the liver of juvenile, castrated, and testosterone-treated rats. *Bio. Chem.* **368**:315–321.

Manley, D. G. 1985. Managing the Japanese beetle on tobacco in South Carolina. *J. Agric. Entomol.* **2**:398–399.

Manning, J. T., Martin, S., Trivers, R. L., and Soler, M. 2002. 2nd to 4th digit ratio and offspring sex ratio. *J. Theor. Biol.* **217**:93–95.

Margolis, R. L., McGinnis, M. G., Rosenblatt, A., and Ross, C. A. 1999. Trinucleotide repeat expansion and neuropsychiatric disease. *Arch. Gen. Psych.* **56**:1019–1031.

Marshall, D. J., Steinberg, P. D., and Evans, J. P. 2004. The early sperm gets the good egg: mating order effects in free spawners. *Proc. R. Soc. Lond. B* **271**:1585–1589.

Martin, L. J., Mahaney, M. C., Bronikowski, A. M., Dee Carey, K., Dyke, B., and Comuzzie, A. G. 2002. Lifespan in captive baboons is heritable. *Mech. Ageing Dev.* **123**:1461–1467.

Martinez, J., Dugaiczyk, L. J., Zielinski, R., and Dugaiczyk, A. 2001. Human genetic disorders, a phylogenetic perspective. *J. Mol. Biol.* **308**:587–596.

Maruyama, J., and Seno, H. 1999. Mathematical modelling for intra-specific brood-parasitism: Coexistence between parasite and non-parasite. *Math. Biosci.* **156**:315–338.

Mather, K., and Jinks, J. L. 1982. *Biometrical Genetics*, 3rd edn. Chapman and Hall, London.

Mayeaux, D. J., Mason, W. A., and Mendoza, S. P. 2002. Developmental changes in responsiveness to parents and unfamiliar adults in a monogamous monkey (*Callicebus moloch*). *Am. J. Primatol.* **58**:71–89.

Maynard Smith, J. 1974. The theory of games and the evolution of animal conflicts. *J. Theor. Biol.* **47**:209–221.

Maynard Smith, J. 1999. Conflict and cooperation in human societies. In: L. Keller, L. (ed.), *Levels of Selection in Evolution*, pp. 197–208, Princeton University Press, Princeton.

Maynard Smith, J., and Price, G. R. 1973. The logic of animal conflict. *Nature* **246**:15–18.

Mayr, E. 1963. *Animal Species and Evolution*. Harvard University Press, Cambridge.

Mazur, J. E. 2002. *Learning and Behavior*, 5th edn. Prentice Hall, Upper Saddle River.

McCleery, R. G. 1978. Optimal behaviour sequences and decision-making. In: J. R. Krebs, and N. B. Davies (eds.), *Behavioural Ecology: An Evolutionary Approach*, pp. 377–410, Sinauer, Sunderland.

McGarty, C., and Turner, J. C. 1992. The effects of categorization on social judgment. *Brit. J. Soc. Psychol.* **31**:253–268.

McGrew, W. C. 1998. Culture in nonhuman primates. *Ann. Rev. Anthropol.* **27**:301–328.

McNab, B. K. 1980. Assessment strategy and the evolution of fighting behavior. *J. Theor. Biol.* **47**:223–243.

McRae, S. B. 1997. A rise in nest predation enhances the frequency of intraspecific brood parasitism in a moorhen population. *J. Anim. Ecol.* **66**:143–153.

Melo, L., Mendes Pontes, A. R., and Monteiro da Cruz, M. A. 2003. Infanticide and cannibalism in wild common marmosets. *Folia Primatol.* **74**:48–50.

Ménard, N. 2002. Ecological plasticity of Barbara macaques (*Macaca sylvanus*). *Evol. Anthropol.* Supp. **1**:95–100.

Mendl, M., Randle, K., and Pope, S. 2002. Young female pigs can discriminate individual differences in odours from conspecific urine. *Anim. Behav.* **64**:97–101.

Miethe, T. D., and McCorkel, R. 1998. *Crime Profiles: The Anatomy of Dangerous Persons, Places, and Situations*. Roxbury Publishing Company, Los Angeles.

Milinski, M., and Wedekind, C. 2001. Evidence for MHC-correlated perfume preferences in humans. *Behav. Ecol.* **12**:140–149.

Miller, A. H. 1956. Ecologic factors that accelerate formation of races and species of terrestrial vertebratesg *Evolution* **10**:262–277.

Miller, G. 2000. *The Mating Mind*. Anchor Books, New York.

Miller, G. 2003. Hungry ewes deliver offspring early. *Science* **300**:561–562.

Miller, G. F. 1997. Protean primates: the evolution of adaptive unpredictability in competition and courtship. In: A. Whiten, and R. W. Byrne (eds.), *Machiavellian Intelligence II: Extensions and Evaluations*, pp. 312–340, Cambridge University Press, Cambridge.

Milton, K. 1980. *The Foraging Strategy of Howler Monkeys*. Columbia University Press, New York.

Milton, K. 1982. Dietary quality and demographic regulation in a howler monkey population. In: E. G. Leigh, Jr., A. S. Rand, and D. M. Windsor (eds.), *The Ecology of a Tropical Forest: Seasonal Rhythms and Long-Term Changes*, pp. 273–289. The Smithsonian Institution Press, Washington.

Mitani, J. C., and Watts, D. 1997. The evolution of non-maternal caretaking among anthropoid primates: do helpers help? *Behav. Ecol. Sociobiol.* **40**:213–220.

Mitchell, A. H. 1994. *Ecology of Hose's Langur, Presbytis hosei, in Mixed Logged and Unlogged Dipterocarp Forest of Northeast Borneo.* Ph.D. Dissertation, Yale University (Unpublished).

Mohtashemi, M., and Mui, L. 2003. Evolution of indirect reciprocity by social information: the role of trust and reputation in evolution of altruism. *J. Theor. Biol.* **223**:523–531.

Moore, A. J., Gowaty, P. A., Wallin, W. G., and Moore, P. J. 2001. Sexual conflict and the evolution of female mate choice and male social dominance. *Proc. R. Soc. Lond. B* **268**:517–523.

Moran, N. A. 1992. The evolutionary maintenance of alternative phenotypes. *Am. Nat.* **139**:971–989.

Morton, C. C. 11 March 2003. In the animal kingdom, female promiscuity may be more the rule than the exception. *The Boston Globe*, C1.

Mousseau, T. A., and Fox, C. W. 1998a. *Maternal Effects as Adaptations.* Oxford University Press, New York.

Mousseau, T. A., and Fox, C. W. 1998b. The adaptive significance of maternal effects. *Trends Ecol. Evol.* **13**:403–407.

Muller, A. E., and Thalmann, U. 2000. Origin and evolution of primate social organisation: a reconstruction. *Biol. Rev. Camb. Philos. Soc.* **75**: 405–435.

Myers, N., Mittermeier, R. A., Mittermeier, C. G., da Fonseca, G. A. B., and Kent, J. 2000. Biodiversity hotspots for conservation priorities. *Nature* **403**:853–858.

Neff, B. D. 2001. Infidelity as a transaction between social mates. *Trends Ecol. Evol.* **16**:175.

Neff, B. D. 2003. Decisions about parental care in response to perceived paternity. *Nature* **422**:716–719.

Neff, B. D., and Sherman, P. W. 2002. Decision making and recognition mechanisms. *Proc. R. Soc. Lond. B* **269**:1435–1441.

Neville, M. K. 1972a. The population structure of red howler monkeys (*Alouatta seniculus*) in Trinidad and Venezuela. *Folia Primatol.* **17**:56–86.

Neville, M. K. 1972b. Social relations within troops of red howler monkeys (*Alouatta seniculus*). *Folia Primatol.* **18**:47–77.

Newton-Fisher, N. E. 2002. Relationships of male chimpanzees in the Budongo Forest, Uganda. In: C. Boesch, G. Hohmann, and L. F. Marchant (eds.), *Behavioural Diversity in Chimpanzees and Bonobos*, pp. 124–137, Cambridge University Press, Cambridge.

Nicolson, N. A. 1987. Infants, mothers, and other females. In: B. B. Smuts, D. L. Cheney, R. M. Seyfarth, R. W. Wrangham, and T. T. Struhsaker (eds.), *Primate Societies*, pp. 330–342, The University of Chicago Press, Chicago.

Nunn, C. L. 1999. The number of males in primate social groups: a comparative test of the socioecological model. *Behav. Ecol. Sociobiol.* **46**:1–13.

Nunn, C. L. 2000. Maternal recognition of infant calls in ring-tailed lemurs. *Folia Primatol.* **71**:142–146.

Nunn, C. L. 2003. Comparative and theoretical approaches to studying sexual selection in primates. In: C. B. Jones (ed.), *Sexual Selection and Reproductive Competition in Primates: New Perspectives and Directions*, pp. 539–613, American Society of Primatologists, Norman. Nunn, C. L., and Pereira, M. E. 2000. Group histories and offspring sex ratios in ringtailed lemurs (Lemur catta). *Behav. Ecol. Sociobiol.* **48**:18–28.

Nur, N., and Hasson, O. 1984. Phenotypic plasticity and the handicap principle. *J. Theor. Biol.* **110**:275–295.

O'Donnell, S. 1997. How parasites can promote the expression of social behavior in their hosts. *Proc. R. Soc. Lond. Series B* **264**:689–694.

Oda, R. 1998. The responses of Verreaux's sifakas to anti-predator alarm calls given by sympatric ring-tailed lemurs. *Folia Primatol.* **69**:357–360.

Odum, E. P. 1971. *Fundamentals of Ecology*, 2nd edn. W. B. Saunders Co., Philadelphia.

Oppenheimer, J. R. 1973. Social and communicatory behavior in the *Cebus* monkey. In: C. R. Carpenter (ed.), *Behavioral Regulators of Behavior*, pp. 251–271, Associated University Presses, Cranbury.

Ord, T. J., and Evans, C. S. 2003. Display rate and opponent assessment in the Jacky Dragon (*Amphibolurus muricatus*): An experimental analysis. *Behaviour* **140**: 1495–1508.

Ostro, L. E. T., Silver, S. C., Koontz, F. W., and Koontz, T. P. 2000. Habitat selection by translocated black howler monkeys in Belize. *Anim. Conserv.* **3**:175–181.

Otte, D. 1975. On the role of intraspecific deception. *Am. Nat.* **109**:239–242.

Padilla, D. K., and Adolph, S. C. 1996. Plastic inducible morphologies are not always adaptive: the importance of time delays in a stochastic environment. *Evol. Ecol.* **10**:105–117.

Page, R. E., Jr., and Erber, J. 2002. Levels of behavioral organization and the evolution of division of labor. *Naturwissenschaften* **89**:91–106.

Pagel, M. 2003. Polygamy and parenting. *Nature* **424**:23–24.

Palleroni, A., and Hauser, M. 2003. Experience-dependent plasticity for auditory processing in a raptor. *Science* **299**:1195.

Pallier, C., Bosch, L., and Sebastian-Galles, N. 1997. A limit on behavioral plasticity in speech perception. *Cognition* **64**:9–17.

Palombit, R. 1992. A preliminary study of vocal communication in wild long-tailed macaques (*Macaca fascicularis*). I. Vocal repertoire and call emission. *Int. J. Primatol.* **13**:143–182.

Palombit, R. A. 2003. Male infanticide in wild savanna baboons: adaptive significance and intraspecific variation. In: C. B. Jones (ed.), *Sexual Selection and Reproductive Competition in Primates: New Perspectives and Directions*, pp. 367–411, American Society of Primatologists, Norman.

Panchanathan, K., and Boyd, R. 2003. A tale of two defectors: the importance of standing for evolution of indirect reciprocity. *J. Theor. Biol.* **224**:115–126.

Parker, G. A. 1974. Assessment strategy and the evolution of fighting behaviour. *J. Theor. Biol.* **47**:223–243.

Parker, G. A. 1979. Sexual selection and sexual conflict. In: M. S. Blum, and N. A. Blum (eds.), *Sexual Selection and Reproductive Competition in Insects*, pp. 123–166, Academic Press, New York.

Parker, J. D., and Rissing, S. W. 2002. Molecular evidence for the origin of workerless social parasites in the ant genus *Pogonomyrmex*. *Evolution* **56**:2017–2028.

Parr, L. A., and de Waal, F. B. M. 1999. Visual kin recognition in chimpanzees. *Nature* **399**:647–648.

Parr, L. A., Matheson, M. D., Bernstein, I. S., and de Waal, F. B. M. 1997. Grooming down the hierarchy: Allogrooming in captive brown capuchin monkeys, *Cebus apella*. *Anim. Behav.* **54**:391–397.

Parthasarathy, H. 2002. Plasticity and the older owl. *Nature* **419**:258–259.

Partridge, L., and Harvey, P. H. 1988. The ecological context of life history evolution. *Science* **241**:1449–1454.

Payne, R. J. H., and Pagel, M. 1996a. Escalation and time costs in displays of endurance. *J. Theor. Biol.* **183**:185–193.

Payne, R. J. H., and Pagel, M. 1996b. When is false modesty a false economy? An optimality model of escalating signals. *Proc. R. Soc. Lond. B* **263**:1545–1550.

Payne, R. J. H., and Pagel, M. 1997. Why do animals repeat displays? *Anim. Behav.* **54**:109–119.

Pelli, D. G., Farell, B., and Moore, D. C. 2003. The remarkable inefficiency of work recognition. *Nature* **423**:752–756.

Pepper, J. W., and Smuts, B. B. 2002. A mechanism for the evolution of altruism among non-kin: Positive assortment through environmental feedback. *Am. Nat.* **160**:205–213.

Perez-Tomé, J. M. and Toro, M. A. 1982. Competition of similar and non-similar genotypes. *Science* **299**:153–154.

Perry, S. 2003. Conclusions and research agendas. In: D. M. Fragaszy, and S. Perry (eds.), *The Biology of Traditions: Models and Evidence*, pp. 426–440, Cambridge University Press, Cambridge.

Piersma, T., and Drent, J. 2003. Phenotypic flexibility and the evolution of organismal design. *Trends Ecol. Evol.* **18**:228–233.

Pigliucci, M. 2001. *Phenotypic Plasticity: Beyond Nature and Nurture.* The Johns Hopkins University Press, Baltimore.

Pimm, S. L., and Raven, P. 2000. Extinction by numbers. *Nature* **403**:843–845.

Plaistow, S. J., Johnstone, R. A., Colegrave, N., and Spencer, M. 2004. Evolution of alternative mating tactics: Conditional versus mixed strategies. *Behav. Ecol.* **15**:534–543.

Pomiankowski, A. 1999. Intragenomic conflict. In: L. Keller (ed.), *Levels of Selection in Evolution*, pp. 121–152, Princeton University Press, Princeton.

Pope, T. R. 1992. The influence of dispersal patterns and mating system on genetic differentiation within and between populations of the red howler monkey (*Alouatta seniculus*). *Evolution* **46**:1112–1128.

Popp, J. L., and DeVore, I. 1979. Aggressive competition and social dominance theory: synopsis. In: D. A. Hamburg, and E. R. McCown (eds.), *The Great Apes*, pp. 317–340, Benjamin/Cummings, Reading.

Porter, L. M. 2001. Social organization, reproduction and rearing strategies of *Callimico goeldii*: new clues from the wild. *Folia Primatol.* **72**:69–79.

Porter, P. G., and Blaustein, A. R. 1989. Mechanisms and ecological correlates of kin recognition. *Sci. Prog.* **73**:53–66.

Poulin, R. 2003. Phenotypic manipulation and parasite-mediated host evolution. In: A. Legakis, S. Sfenthourakis, R. Polymeni, and M. Thessalou-Legaki (eds.), *The New Panorama of Animal Evolution*, pp. 205–212, Pensoft Publishers, Sofia.

Poulin, R., and Thomas, F. 1999. Phenotypic variability induced by parasites: extent and evolutionary implications. *Parasitol. Today* **15**:28–32.

Pound, N. 2002. Male interest in visual cues of sperm competition risk. *Evol. Hum. Behav.* **23**:443–466.

Premack, D. 2004. Is language the key to human intelligence? *Science* **303**:318–320.

Preti, G., Wysocki, C. J., Barnhart, K. T., Sondheimer, S. J., and Leyden, J. J. 2003. Male axillary extracts contain pheromones that affect pulsatile secretion of luteinizing hormone and mood in women recipients. *Biol. Reprod.* **68**:2107–2113.

Preutz, J. D., and Isbell, L. A. 2000. Correlations of food distribution and patch size with agonistic interactions in female vervets (*Chlorocebus aethiops*) and patas monkeys (*Erythrocebus patas*) living in simple habitats. *Behav. Ecol. Sociobiol.* **49**:38–47.

Proulx, S. R. 1999. Mating systems and the evolution of niche breadth. *Am. Nat.* **154**:89–98.

Proulx, S. R., Day, T., and Rowe, L. 2002. Older males signal more reliably. *Proc. R. Soc. Lond. B* **269**:2291–2299.

Pulliam, R., and Caraco, T. 1984. Living in groups: Is there an optimal group size? In: J. R. Krebs, and N. B. Davies (eds.), *Behavioural Ecology: An Evolutionary Approach*, pp. 122–147, Sinauer Associates, Inc., Sunderland.

Pusey, A. E., and Packer, C. 1987. Dispersal and philopatry. In: B. B. Smuts, D. L. Cheney, R. M. Seyfarth, R. W. Wrangham, and T. T. Struhsaker (eds.), *Primate Societies*, pp. 250–266, The University of Chicago Press, Chicago.

Queller, D. C. 1997. Why do females care more than males? *Proc. R. Soc. Lond. B* **264**:1555–1557.

Queller, D. C., Zacchi, F., Cervo, R., Turillazzi, S., Henshaw, M. T., Santorelli, L. A., and Strassmann, J. E. 2000. Unrelated helpers in a social insect. *Nature* **405**:784–787.

Qvarnstrom, A., Part, T., and Sheldon, B. C. 2000. Adaptive plasticity in mate preferences linked to differences in reproductive effort. *Nature* **405**:344–347.

Radespiel, U., and Zimmermann, E. 2001. Female dominance in captive gray mouse lemurs. *Am. J. Primatol.* **54**:181–192.

Radespiel, U., Cepok, S., Zietemann, V., and Zimmermann, E. 1998. Sex-specific usage patterns of sleeping sites in gray mouse lemurs (*Microcebus murinus*) in northwestern Madagascar. *Am. J. Primatol.* **46**:77–84.

Randall, J. A., Hekkala, E. R., Cooper, L. D., and Barfield, J. 2002. Familiarity and flexible mating strategies of a solitary rodent Dipodomys ingens. *Anim. Behav.* **64**:11–21.

Reader, S. M., and Macdonald, K. 2003. Environmental variability and primate behavioural flexibility. In: S. M. Reader, and K. N. Laland (eds.), *Animal Innovation*, pp. 83–116, Oxford University Press, Oxford.

Reeder, D.-A. M. 2003. The potential for cryptic female choice in primates: behavioral, anatomical, and physiological considerations. In: C. B. Jones (ed.), *Sexual Selection and Reproductive Competition in Primates: New Perspectives and Directions*, pp. 255–303, American Society of Primatologists, Norman.

Reeve, H. K. 2001. In search of unified theories in sociobiology: help from social wasps. In: L. A. Dugatkin (ed.), *Model Systems in Behavioral Ecology: Integrating Conceptual, Theoretical, and Empirical Approaches*, pp. 57–71, Princeton University Press, Princeton.

Reeve, H. K., and Emlen, S. T. 2000. Reproductive skew and group size: an *N*-person staying incentive model. *Behav. Ecol.* **11**:640–647.

Relyea, R. A. 2002. Costs of phenotypic plasticity. *Am. Nat.* **159**:272–282.

Rice, W. R. 2000. Dangerous liaisons. *Proc. Nat. Acad. Scis. USA* **97**:12953–12955.

Rice, W. R., and Chippendale, A. K. 2001. Intersexual ontogenetic conflict. *J. Evol. Biol.* **14**:685–693.

Rice, W. R., and Holland, B. 1997. The enemies within: intergenomic conflict, interlocus contest evolution (ICE), and the intraspecific Red Queen. *Behav. Ecol. Sociobiol.* **41**:1–10.

Robinson, G. E. 1999. Integrative animal behaviour and sociogenomics. *Trends Ecol. Evol.* **14**:202–205.

Robinson, G. E. 2002. Sociogenomics takes flight. *Science* **297**:204–205.

Robinson, G. E., and Ben-Shahar, Y. 2002. Social behavior and comparative genomics: new genes or new gene regulation? *Genes, Brain and Behav.* **1**:197–203.

Robinson, G. E., Fahrbach, S. E., and Winston, M. L. 1997. Insect societies and the molecular biology of social behavior. *Bioessays* **19**:1099–1108.

Rodman, P. S., and Mitani, J. C. 1987. Orangutans: Sexual dimorphism in a solitary species. In: B. B. Smuts, D. L. Cheney, R. M. Seyfarth, R. W. Wrangham, and T. T. Struhsaker (eds.), *Primate Societies*, pp. 146–154, The University of Chicago Press, Chicago.

Rodriguez, I., Greer, C. A., Mok, M. Y., and Mombaerts, P. 2000. A putative pheromone receptor gene expressed in human olfactory mucosa. *Nat. Genet.* **26**:18–19.

Root, R. B. 1967. The niche exploitation pattern of the blue-gray gnatcatcher. *Ecol. Monogr.* **37**:331–349.

Rosch, E. 1975. Cognitive representations of semantic categories. *J. Exp. Psychol. Gen.* **104**:192–233.

Ross, C., and MacLarnon, A. 2000. The evolution of non-maternal care in anthropoid primates: a test of the hypotheses. *Folia Primatol.* **71**:93–113.

Roughgarden, J. 1979. *Theory of Population Genetics and Evolutionary Ecology: An Introduction*. Macmillan Publishing Co., Inc., New York.

Roughgarden, J. 1983. The theory of coevolution. In: D. J. Futuyma, and M. Slatkin (eds.), *Coevolution*, pp. 33–64, Sinauer Associates, Sunderland.

Roughgarden, J. 1998. *Primer of Ecological Theory*. Prentice Hall, Upper Saddle River.

Roughgarden, J. 2004. *Evolution's Rainbow*. University of California Press, Berkeley.

Roush, R. S., and Snowdon, C. T. 1999. The effects of social status on food-associated calling behaviour in captive cotton-top tamarins. *Anim. Behav.* **58**:1299–1305.

Royle, N. J., Hartley, I. R., and Parker, G. A. 2002. Sexual conflict reduces offspring fitness in zebra finches. *Nature* **416**:733–736.

Rudran, R. 1979. The demography and social mobility of a red howler (*Alouatta seniculus*) population in Venezuela. In: J. F. Eisenberg (ed.), *Vertebrate Ecology in the Northern Neotropics*, pp. 107–126, Smithsonian Institution Press, Washington.

Runciman, W. G., Maynard Smith, J., and Dunbar, R. I. M. 1996. *Evolution of Social Behaviour Patterns in Primates and Man.* Oxford University Press, Oxford.

Russell, A. F., and Hatchwell, B. J. 2001. Experimental evidence for kin-based helping in a cooperatively breeding vertebrate. *Proc. R. Soc. Lond. B* **268**:2169–
2174.

Russell, A. F., Sharpe, L. L., Brotherton, P. N. M., and Clutton-Brock, T. H. 2003. Cost
minimization by helpers in cooperative vertebrates. *Proc. Nat. Acad. Scis. USA*
100:3333–3338.

Saltzman, W. 2003. Reproductive competition among female common marmosets
(*Callithrix jacchus*): Proximate and ultimate causes. In: C. B. Jones (ed.), *Sexual
Selection and Reproductive Competition in Primates: New Perspectives and Directions*,
pp. 197–229, American Society of Primatologists, Norman.

Sammeta, K. P., and Levins, R. 1970. Genetics and ecology. *Ann. Rev. Genet.* **4**:469–
488.

Savolainen, R., and Vepsäläinen, K. 2003. Sympatric speciation through intraspecific
social parasitism. *Proc. Nat. Acad. Scis. USA* **100**:7169–7174.

Scheiner, S. M. 1993. Genetics and evolution of phenotypic plasticity. *Ann. Rev. Ecol.
Syst.* **24**:35–68.

Schlichting, C. D., and Pigliucci, M. 1998. *Phenotypic Evolution: A Reaction Norm Perspective*. Sinauer Associates, Inc., Sunderland.

Schoener, T. W. 1971. Theory of feeding strategies. *Ann. Rev. Ecol. Syst.* **2**:369–404.

Schoener, T. W. 1974. Resource partitioning in ecological communities. *Science* **185**:27–
39.

Schulke, O. 2001. Social anti-predator behaviour in a nocturnal lemur. *Folia Primatol.*
72:332–334.

Schwarz, M. P., Silberbauer, L. X., and Hurst, P. S. 1997. Intrinsic and extrinsic fac-
tors associated with social evolution in allodapine bees. In: J. C. Choe, and B. J.
Crespi (eds.), *The Evolution of Social Behavior in Insects and Arachnids*, pp. 333–346,
Cambridge University Press, New York.

Scott, N. J., Jr., Malmgren L. A., and Glander K. E. 1978. Grouping behavior and sex
ratio in mantled howling monkeys. In: D. J. Chivers, and W. Lane-Petter (eds.),
*Proceedings of the Sixth Annual International Congress of the International Primatological
Society*, pp. 183–185, London: Academic Press, London.

Scriver, C. R., and Waters, P. J. 1999. Monogenetic traits are not simple: lessons from
phenylketonuria. *Trends Genet.* **15**:267–272.

Semmann, D., Krambeck, H.-J., and Millinski, M. 2003. Volunteering leads to rock-
paper-scissors dynamics in a public goods game. *Nature* **425**:390–393.

Seyfarth, R. M., Cheney, D. L., and Marler, P. 1980. Monkey responses to three different
alarm calls: Evidence for predator classification and semantic communication.
Science **210**:801–803.

Shahnoor, N., and Jones, C. B. 2003. A brief history of the study of sexual selection
and reproductive competition in primatology. In: C. B. Jones (ed.), *Sexual Selection
and Reproductive Competition in Primates: New Perspectives and Directions*, pp. 1–43.
American Society of Primatologists, Norman.

Schaffner, C. M., and Caine, N. G. 2000. The peacefulness of cooperatively breeding
primates. In: F. Aureli, and F. B. M. de Waal (eds.), *Natural Conflict Resolution*,
pp. 155–169, University of California Press, Berkeley.

Shaw, E., and Innes, K. 1980. The "infantilization" of a cichlid fish. *Dev. Psychobiol.*
13:131–139.

Shellman-Reeve, J. S., and Reeve, H. K. 2000. Extra-pair paternity as the result of reproductive transactions between paired mates. *Proc. R. Soc. Lond. B* **267**:2543–2546.

Sherman, G., and Visscher, P. K. 2002. Honeybee colonies achieve fitness through dancing. *Nature* **419**:920–922.

Sherman, P. W., and Neff, B. D. 2003. Father knows best. *Nature* **425**:136–137.

Sherman, P. W., Jarvis, J. U., and Alexander, R. D. (eds.) 1991. *The Biology of the Naked Mole-Rat*. Princeton University Press, Princeton.

Sherman, P. W., Lacey, E. A., Reeve, H. K., and Keller, L. 1995. The eusociality continuum. *Behav. Ecol.* **6**:102–108.

Shine, R., Phillips, B., Wayne, H., LeMaster, M., and Mason, R. T. 2001. Benefits of female mimicry in snakes. *Nature* **414**:267.

Shuster, S. M., and Wade, M. J. 2003. *Mating Systems and Strategies*. Princeton University Press, Princeton.

Sidman, M., and Tailby, W. 1982. Conditional discrimination versus matching to sample: An extension of the testing paradigm. *J. Exp. Anal. Behav.* **37**:5–22.

Silk, J. B. 1987. Social behavior in evolutionary perspective. In: B. B. Smuts, D. L. Cheney, R. M. Seyfarth, R. W. Wrangham, and T. T. Struhsaker (eds.), *Primate Societies*, pp. 318–329, The University of Chicago Press, Chicago.

Silk, J. B. 1993. The evolution of social conflict among female primates. In: W. A. Mason, and S. P. Mendoza, (eds.), *Primate Social Conflict*, pp. 49–83, SUNY Press, Albany.

Silk, J. B. 2003. Cooperation without counting: The puzzle of friendship. In: P. Hammerstein (ed.), *Genetic and Cultural Evolution of Cooperation*, pp. 37–54, The MIT Press, Cambridge.

Silk, J. B., Alberts, S. C., and Altmann, J. 2003. Social bonds of female baboons enhance infant survival. *Science* **302**:1231–1234.

Silva, A. J., Smith, A. M., and Giese, K. P. 1997. Gene targeting and the biology of learning and memory. *Ann. Rev. Genet.* **31**:527–546.

Silver, S. C., and Marsh, L. K. 2003. Dietary flexibility, behavioral plasticity, and survival in fragments: lessons from translocated howlers. In: L. K. Marsh (ed.), *Primates in Fragments: Ecology and Conservation*, pp. 251–266, Kluwer, New York.

Sinervo, B., and Clobert, J. 2003. Morphs, dispersal behavior, genetic similarity, and the evolution of cooperation. *Science* **300**:1949–1951.

Slatkin, M. 1974. Competition and regional coexistence. *Ecology* **55**:128–134.

Slobodkin, L. B. 1968. Toward a predictive theory of evolution. In: R. C. Lewontin (ed.), *Population Biology and Evolution*, pp. 187–205, Syracuse University Press, New York.

Slobodkin, L. B., and Rapoport, A. 1974. An optimal strategy of evolution. *Quart. Rev. Biol.* **49**:181–200.

Smith, M. H., and Joule, J. (eds.) 1981. *Mammalian Population Genetics*. University of Georgia Press, Athens.

Smith, T., and Polacheck, T. 1981. Reexamination of the life table for northern fur seals with implications about population regulation mechanisms. In: C. W. Fowler, and T. D. Smith (eds.), *Dynamics of Large Mammal Populations*, pp. 99–120, John Wiley and Sons, New York.

Smuts, B. B. 1985. *Sex and Friendship in Baboons*. Aldine, New York.

Smuts, B. B. 1987a. Sexual competition and mate choice. In: B. B. Smuts, D. L. Cheney, R. M. Seyfarth, R. W. Wrangham, and T. T. Struhsaker (eds.), *Primate Societies*, pp. 385–399, The University of Chicago Press, Chicago.

Smuts, B. B. 1987b. Gender, aggression, and influence. In: B. B. Smuts, D. L. Cheney, R. M. Seyfarth, R. W. Wrangham, and T. T. Struhsaker (eds.), *Primate Societies*, pp. 400–412, The University of Chicago Press, Chicago.

Smuts, B. B., and Smuts, R. W. 1993. Male aggression and sexual coercion of females in nonhuman primates and other mammals: evidence and theoretical implications. *Adv. Stud. Behav.* **22**:1–63.

Smuts, B. B., and Watanabe, J. M. 1990. Social relationships and ritualized greetings in adult male baboons (*Papio cynocephalus anubis*). *Int. J. Primatol.* **11**:147–172.

Smuts, B. B., Cheney, D. L., Seyfarth, R. M., Wrangham, R. W., and Struhsaker, T. T. (eds.) 1987. *Primate Societies*. The University of Chicago Press, Chicago.

Snowdon, C. C. 1997. The "nature" of sex differences: Myths of male and female. In: P. A. Gowaty (ed.), *Feminism and Evolutionary Biology: Boundaries, Intersections, and Frontiers*, pp. 276–293, Chapman and Hall, New York.

Soler, M. 2001. Begging behaviour of nestlings and food delivery by parents: The importance of breeding strategy. *Acta Ethol.* **4**:59–63.

Solomon, N. G., and French, J. A. 1997. *Cooperative Breeding in Mammals*. Cambridge University Press, Cambridge.

Soltis, J., Mitsunaga, F., Shimizu, K., Yanagihara, Y., and Nozaki, M. 1997a. Sexual selection in Japanese macaques I: female mate choice or male sexual coercion? *Anim. Behav.* **54**:725–736.

Soltis, J., Mitsunaga, F., Shimizu, K., Nozaki, M., Yanagihara, Y., Domingo-Roura, X., and Takenaka, O. 1997b. Sexual selection in Japanese macaques II: female mate choice and male–male competition. *Anim. Behav.* **54**:737–746.

Soltis, J., Mitsunaga, F., Shimizu, K., Yanagihara, Y., and Nozaki, M. 1999. Female mating strategy in an enclosed group of Japanese macaques. *Am. J. Primatol.* **47**:263–278.

Sonsino, D., and Mandelbaum, M. 2001. On preference for flexibility and complexity aversion: experimental evidence. *Theory Decis.* **51**:197–216.

Spencer-Booth, Y. 1970. The relationships between mammalian young and conspecifics other than mothers and peers: A review. *Adv. Stud. Behav.* **3**:119–194.

Spitze, K., and Sadler, T. D. 1996. Evolution of a generalist genotype: multivariate analysis of the adaptiveness of phenotypic plasticity. *Am. Nat.* **148**:S108–S123.

Staddon, J. E. R. 1983. *Adaptive Behavior and Learning*. Cambridge University Press, New York.

Stearns, S. C. 1976. Life-history tactics: a review of the ideas. *Quart. Rev. Biol.*:**51**:3–47.

Stearns, S. C. 1992. *The Evolution of Life Histories*. Oxford University Press, Oxford.

Stearns, S. C. 2002. Progress on canalization. *Proc. Nat. Acad. Scis. USA* **99**:10229–10230.

Stearns, S. C. 2003. Safeguards and spurs. *Nature* **424**:501–504.

Stearns, S. C., de Jong, G., and Newman, B. 1991. The effects of phenotypic plasticity on genetic correlations. *Trends Ecol. Evol.* **6**:122–126.

Sterck, E. H. 1997. Determinants of female dispersal in Thomas langurs. *Am. J. Primatol.* **42**:179–198.

Sterck, E. H. M. 1998. Female dispersal, social organization, and infanticide in langurs: are they linked to human disturbance? *Am. J. Primatol.* **44**:235–254.

Sterck, E. H. M. 1999. Variation in langur social organization in relation to the socioecological model, human habitat alteration, and phylogenetic constraints. *Primates* **40**:199–213.

Sterck, E. H. M., Watts, D. P., and van Schaik, C. P. 1997. The evolution of female social relationships in nonhuman primates. *Behav. Ecol. Sociobiol.* **4**:291–309.

Stern, K., and McClintock, M. K. 1998. Regulation of ovulation by human pheromones. *Nature* **392**:177–179.

Stevens, J. R. 2003. The selfish nature of generosity: Harassment and food sharing in primates. *Proc. R. Soc. Lond. B* **271**:451–456.

Stokke, B. G., Moksnes, A., and Roskaft, E. 2002. Obligate brood parasites as selective agents for evolution of egg appearance in passerine birds. *Int. J. Org. Evol.* **56**:199–205.

Strier, K. B. 1992. Atelinae adaptations: behavioral strategies and ecological constraints. *Am. J. Phys. Anthropol.* **88**:515–524.

Strier, K. B. 2000. From binding brotherhoods to short-term sovereignty: The dilemma of male Cebidae. In: P. M. Kappeler (ed.), *Primate Males: Causes and Consequences of Variation in Group Composition*, pp. 72–83, Cambridge University Press, Cambridge.

Strier, K. B. 2003. Demography and the temporal scale of sexual selection. In: C. B. Jones (ed.), *Sexual Selection and Reproductive Competition in Primates: New Perspectives and Directions*, pp. 45–63, American Society of Primatologists, Norman.

Strier, K. B., Dib, L. T., and Figueira, J. E. C. 2002. Social dynamics of male muriquis (*Brachyteles arachnoides hypoxanthus*). *Behaviour* **139**:315–342.

Strier, K. B., Mendes, S. L., and Santos, R. R. 2001. Timing of births in sympatric brown howler monkeys (*Alouatta fusca clamitans*) and northern muriquis (*Brachyteles arachnoides hypoxanthus*). *Am. J. Primatol.* **55**:87–100.

Subramanian, S., and Kumar, S. 2003. Neutral substitutions occur at a faster rate in exons than in noncoding DNA in primate genomes. *Genome Res.* **13**:838–844.

Sultan, S. E., and Spencer, H. G. 2002. Metapopulation structure favors plasticity over local adaptation. *Am. Nat.* **160**:271–283.

Summers, K., McKeon, S., Sellars, J., Keusenkothen, M., Morris, J., Gloeckner, D., Pressley, C., Price, B., and Snow, H. 2003. Parasitic exploitation as an engine of diversity. *Biol. Rev.* **78**:639–675.

Sundstrom, H., Webster, M. T., and Ellegren, H. 2003. Is the rate of insertion and deletion mutation male biased? Molecular evolutionary analysis of avian and primate sex chromosome sequences. *Genetics* **164**:259–268.

Swedell, L. 2002. Ranging behavior, group size and behavioral flexibility in Ethiopian Hamadryas baboons (*Papio hamadryas hamadryas*). *Folia Primatol.* **73**:95–103.

Taber, R. D., and Dasmann, R. F. 1957. The dynamics of three natural populations of deer (*Odocoileus hemionus columbianus*). *Ecology* **38**:233–246.

Taborsky, M. 1994. Sneakers, sattelites, and helpers: parasitic and cooperative behavior in fish reproduction. *Adv. Stud. Behav.* **23**:1–100.

Thierry, B. 2000. Covariation of conflict management patterns across macaque species. In: F. Aureli, and F. B. M. de Waal (eds.), *Natural Conflict Resolution*, pp. 106–128, University of California Press, Berkeley.

Thorington, R. W., Jr., Rudran, R., and Mack, D. 1979. Sexual dimorphism of *Alouatta seniculus* and observations on capture techniques. In: J. F. Eisenberg (ed.),

Vertebrate Ecology in the Northern Neotropics, pp. 97–106, Smithsonian Institution Press, Washington.

Thornhill, R., and Palmer, C. T. 2000. *A Natural History of Rape: Biological Bases of Sexual Coercion*. The MIT Press, Cambridge.

Tilman, D. 1999. Diversity by default. *Science* **283**:495–496.

Tinbergen, N. 1952. "Derived" activities: their causation, biological significance, origin, and emancipation during evolution. *Quart. Rev. Biol.* **27**:1–32.

Tofilski, A. 2002. Influence of age polyethism on longevity of workers in social insects. *Behav. Ecol. Sociobiol.* **51**:234–237.

Tolkamp, B. J., Emmans, G. C., Yearsley, J., and Kyriazakis, I. 2002. Optimization of short-term animal behaviour and the currency of time. *Anim. Behav.* **64**:945–953.

Tomasello, M., and Call, J. 1997. *Primate Cognition*. Oxford University Press, Oxford.

Traniello, J. F., and Rosengaus, R. B. 1997. Ecology, evolution and division of labour in social insects. *Anim. Behav.* **53**:209–213.

Travis, J. 1994. Ecological genetics of life-history traits: Variation and its evolutionary significance. In: L. A. Real (ed.), *Ecological Genetics*, pp. 171–204, Princeton University Press, Princeton.

Treves, A., and Chapman, C. A. 1996. Conspecific threat, predation avoidance, and resource defense: implications for grouping in langurs. *Behav. Ecol. Sociobiol.* **39**:43–53.

Trivers, R. L. 1971. The evolution of reciprocal altruism. *Quart. Rev. Biol.* **46**:35–57.

Trivers, R. L. 1972. Parental investment and sexual selection. In: B. Campbell (ed.), *Sexual Selection and the Descent of Man, 1871–1971*, pp. 136–179, Aldine, New York.

Trivers, R. L. 1974. Parent-offspring conflict. *Am. Zool.* **14**:249–264.

Trivers, R. L. 1985. *Social Evolution*. The Benjamin/Cummings Publishing Company, Inc., Menlo Park.

van Lawick-Goodall, J. 1968. The behaviour of free ranging chimpanzees in the Gombe stream area. *Anim. Behav. Monogr.* **1**:161–311.

van Lawick-Goodall, J. 1970. Tool-using in primates and other vertebrates. *Adv. Stud. Behav.* **3**:195–249.

van Schaik, C. P. 1989. The ecology of social relationships amongst female primates. In: V. Standen, and R. A. Foley (eds.), *Comparative Socioecology: The Behavioural Ecology of Humans and Other Mammals*, pp. 195–218, Blackwell Scientific Publications, Oxford.

van Schaik, C. P., and Kappeler, P. M. 1997. Infanticide risk and the evolution of male-female association in primates. *Proc. R. Soc. Lond. B* **264**:1687–1694.

van Schaik, C. P., and Janson, C. H. (eds.) 2000. *Infanticide by Males and Its Implications*. Cambridge University Press, Cambridge.

van Schaik, C. P., Ancrenaz, M., Borgen, G., Galdikas, B., Knott, C. D., Singleton, I., Suzuki, A., Utami, S. S., and Merrill, M. 2003. Orangutan cultures and the evolution of material culture. *Science* **299**:102–105.

Van Tienderen, P. H. 1991. Evolution of generalists and specialists in spatially heterogeneous environments. *Evolution* **45**:1317–1331.

Van Tienderen, P. H., and Koelewijn, H. P. 1994. Selection on reaction norms, genetic correlations and constraints. *Genet. Res.* **64**:115–125.

Vandenbergh, J. G. 1983. The role of hormones in synchronizing mammalian reproductive behavior. In: J. F. Eisenberg., and D. G. Kleiman (eds.), *Advances in the*

Study of Mammalian Behavior, pp. 95–112, The American Society of Mammalogists, Shippensburg.

Vasey, P. L. 1995. Homosexual behavior in primates: a review of evidence and theory. *Int. J. Primatol.* **16**:173–204.

Vasey, P. L. 1998. Female choice and inter-sexual competition for female sexual partners in Japanese macaques. *Behaviour* **135**:579–597.

Vasey, P. L. 2000. Skewed sex ratios and female homosexual activity in Japanese macaques: An experimental analysis. *Primates* **41**:17–25.

Vasey, P. L. 2002. Sexual partner preference in female Japanese macaques. *Arch. Sex. Behav.* **31**:51–62.

Vaughan, T. A. 1978. *Mammalogy*. Saunders College, Philadelphia.

Velando, A. 2002. Experimental manipulation of maternal effort produces differential effects in sons and daughters: Implications for adaptive sex ratios in the blue-footed booby. *Behav. Ecol.* **13**:443–449.

Velicer, G. J., and Yu, Y. N. 2003. Evolution of novel cooperative swarming in the bacterium. *Myxococcus xanthus. Nature* **425**:75–78.

Vervaecke, H., Stevens, J., and Van Elsacker, L. 2003. Interfering with others: Female-female reproductive competition in Pan paniscus. In: C. B. Jones (ed.), *Sexual Selection and Reproductive Competition in Primates: New Perspectives and Directions*, pp. 231–253, American Society of Primatologists, Norman.

Vickery, W. L., Brown, J. S., and FitzGerald, G. J. 2003. Spite: Altruism's evil twin. *Oikos* **102**:413–416.

Waage, J. K. 1988. Confusion over residency and the escalation of damselfly territorial disputes. *Anim. Behav.* **36**:586–595.

Wachtmeister, C.-A., and Enquist, M. 2000. The evolution of courtship rituals in monogamous species. *Behav. Ecol.* **11**:405–410.

Wahl, L. M. 2002. Evolving the division of labour: generalists, specialists and task allocation. *J. Theor. Biol.* **219**:371–388.

Walters, J. R., and Seyfarth, R. M. 1987. Conflict and cooperation. In: B. B. Smuts, D. L. Cheney, R. M. Seyfarth, R. W. Wrangham, and T. T. Struhsaker (eds.), *Primate Societies*, pp. 306–317, The University of Chicago Press, Chicago.

Wand, E., and Milton, K. 2003. Intragroup social relationships of male *Alouatta palliata* on Barro Colorado Island, Republic of Panama. *Int. J. Primatol.* **24**:1227–1243.

Wasserman, M. D., and Chapman, C. A. 2003. Determinants of colobus monkey abundance: The importance of food energy, protein and fibre content. *J. Anim. Ecol.* **72**:650–659.

Watson, A., and Moss, R. 1971. Spacing as affected by territorial behavior, habitat and nutrition in red grouse (*Lagopus l. scoticus*). In: A. H. Esser (ed.), *Behavior and Environment: The Use of Space by Animals and Men*, pp. 92–111, Plenum Press, New York.

Watson, P. J., Arnqvist, G., and Stallmann, R. R. 1998. Sexual conflict and the energetic costs of mating and mate choice in water striders. *Am. Nat.* **151**:46–58.

Watson, S., Bingham, W., Stavisky, R., Gray, A., and Fontenot M. B. 2003. Sex ratio bias from the effects of parity on the reproductive characteristics of Garnett's bushbaby: Implications for sexual selection. In: C. B. Jones (ed.), *Sexual Selection and Reproductive Competition in Primates: New Perspectives and Directions*, pp. 173–195, American Society of Primatologists, Norman.

Wauters, L., Matthysen, E., and Dhondt, A. A. 1994. Survival and lifetime reproductive success in dispersing and resident red squirrels. *Behav. Ecol. Sociobiol.* **34**:197–201.

Wcislo, W. T. 1987. The roles of seasonality, host synchrony, and behaviour in the evolutions and distributions of nest parasites in Hymenoptera (Insecta), with special reference to bees (Apoidea). *Biol. Rev.* **62**:515–543.

Wcislo, W. T. 1997. Are behavioral classifications blinders to studying natural variation? In: J. C. Choe, and B. J. Crespi (eds.), *The Evolution of Social Behavior in Insects and Arachnids*, pp. 8–13, Cambridge University Press, Cambridge.

Weaver, I. C. G., Cervoni, N., Champagne, F. A., D'Alessio, A. C., Sharma, S., Secki, J. R., Dymov, S., Szyf, M., and Meaney, M. J. 2004. Epigenetic programming by maternal behavior. *Nat. Neurosci.* **7**:847–854.

Wedekind, C., Seebeck, T., Bettens, F., and Paepke, A. J. 1995. MHC-dependent mate preferences in humans. *Proc. R. Soc. Lond. B* **260**:245–249.

Wells, R. D., and Warren, S. T. 1998. *Genetic Instabilities and Hereditary Neurological Diseases.* Academic Press, San Diego.

West, M. J. 1967. Foundress associations in polistine wasps: Dominance hierarchies and the evolution of social behavior. *Science* **157**:1584–1585.

West, S. A., Murray, M. G., Machado, C. A., Griffin, A. S., and Herre, E. A. 2001. Testing Hamilton's rule with competition between relatives. *Nature* **409**:510–513.

West, S. A., Pen I. and Griffin A. S. 2002. Cooperation and competition between relatives. *Science* **296**:72–75.

West-Eberhard, M. J. 1975. The evolution of social behavior by kin selection. *Quart. Rev. Biol.* **50**:1–33.

West-Eberhard, M. J. 1979. Sexual selection, social competition, and evolution. *Proc. Am. Phil. Soc.* **123**:222–234.

West-Eberhard, M. J. 1983. Sexual selection, social competition, and speciation. *Quart. Rev. Biol.* **58**:155–183.

West-Eberhard, M. J. 1986. Alternative adaptations, speciation, and phylogeny (a review. *Proc. Nat. Acad. Scis. USA* **83**:1388–1392.

West-Eberhard, M. J. 1989. Phenotypic plasticity and the origins of diversity. *Ann. Rev. Ecol. Syst.* **20**:249–278.

West-Eberhard, M. J. 2003. *Developmental Plasticity and Evolution.* Oxford University Press, Oxford.

Whiten, A., and Byrne, R. W. 1988. The manipulation of attention in primate tactical deception. In: R. W. Byrne, and A. Whiten (eds.), *Machiavellian Intelligence: Social Expertise and the Evolution of Intellect in Monkeys, Apes, and Humans*, pp. 211–223, Clarendon Press, Oxford.

Whiten, A., and Byrne, R. W. 1997. *Machiavellian Intelligence II: Extensions and Evaluations.* Cambridge University Press, Cambridge.

Whiten, A., Goodall, J., McGrew, W. C., Nishida, T., Reynolds, V., Sugiyama, Y., Tutin, C. E. G., Wrangham, R. W., and Boesch, C. 1999. Cultures in chimpanzees. *Nature* **399**:682–685.

Whitfield, J. 2002. Nosy neighbours. *Nature* **419**:242–243.

Whitlock, M. C. 1996. The red queen beats the jack-of-all-trades: the limitations on the evolution of phenotypic plasticity and niche breadth. *Am. Nat.* **148**:S65–S77.

Widdig, A., Bercovitch, F. B., Streich, W. J., Sauermann, U., Nurnberg, P., and Krawczak, M. 2004. A longitudinal analysis of reproductive skew in male rhesus macaques. *Proc. R. Soc. Lond. B* **271**:819–826.

Wieczkowski, J. A. 2003. *Aspects of the Ecological Flexibility of the Tana Mangabey (Cercocebus galeritus) in Its Fragmented Habitat, Tana River, Kenya.* Ph.D. Dissertation, University of Georgia (Unpublished).

Wiens, J. A. 2001. The landscape context of dispersal. In: J. Clobert, E. Danchin, A. A. Dhondt, and J. D. Nichols (eds.), *Dispersal*, pp. 96–109, Oxford University Press, Oxford.

Williams, G. C. 1966. *Adaptation and Natural Selection.* Princeton University Press, Princeton.

Wilson, D. S. 1980. *The Natural Selection of Populations and Communities.* The Benjamin/Cummings Publishing Company, Inc., Menlo Park, CA.

Wilson, E. O. 1971. *The Insect Societies.* Harvard University Press, Cambridge.

Wilson, E. O. 1975. *Sociobiology: The New Synthesis.* Harvard University Press, Cambridge.

Wittenberger, J. F. 1980. Group size and polygamy in social mammals. *Am. Nat.* **115**:197–222.

Wolfe, L. D. 1981. Display behavior of three troops of Japanese monkeys (*Macaca fuscata*). *Primates* **22**:24–32.

Wolff, J. O., and Macdonald, D. W. 2004. Promiscuous females protect their offspring. *Trends Ecol. Evol.* **19**:127–134.

Wolfheim, J. H. 1983. *Primates of the World: Distribution, Abundance, and Conservation.* The University of Washington Press, Seattle.

Wourms, J. P. 1972. Developmental biology of annual fishes. III Pre-embryonic and embryonic diapause of variable duration in the eggs of annual fish. *J. Exper. Zool.* **182**:389–414.

Wrangham, R., and Peterson, D. 1996. *Demonic Males: Apes and the Origins of Human Violence.* Houghton Mifflin Company, Boston.

Wrangham, R. W. 1979. On the evolution of ape social systems. *Soc. Sci. Infor.* **18**:334–368.

Wrangham, R. W. 1980. An ecological model of female-bonded primate groups. *Behaviour* **75**:262–300.

Wrangham, R. W. 1987. Evolution of social structure. In: B. B. Smuts, D. L. Cheney, R. M. Seyfarth, R. W. Wrangham, and T. T. Struhsaker (eds.), *Primate Societies*, pp. 282–296, The University of Chicago Press, Chicago.

Wright, W. G. 2000. Neuronal and behavioral plasticity in evolution: Experiments in a model lineage. *BioScience* **50**:883–894.

Wynne, C. D. L. 2004. The perils of anthropomorphism. *Nature* **428**:606.

Yamamoto, J., and Asano, T. 1995. Stimulus equivalence in a chimpanzee (*Pan troglodytes*). *Psychol. Rec.* **45**:3–21.

Yi, S., Ellsworth, D. L., and Li, W. H. 2002. Slow molecular clocks in Old World monkeys, apes, and humans. *Mol. Biol. Evol.* **19**:2191–2198.

Zahavi, A. 1974. Communal nesting by the Arabian babbler. *Ibis* **116**:84–87.

Zahavi, A. 1975. Mate selection—a selection for a handicap. *J. Theor. Biol.* **53**:205–213.

Zahavi, A. 2003. Indirect selection and individual selection in sociobiology: my personal views on theories of social behavior. *Anim. Behav.* **65**:859–863.

Zahavi, A., and Zahavi, A. 1997. *The Handicap Principle.* Oxford University Press, Oxford.

Ziegler, T. E., Savage, A., Scheffler, G., and Snowdon, C. T. 1987. The endocrinology of puberty and reproductive functioning in female cotton-top tamarins (*Saguinus oedipus*) under varying social conditions. *Biol. Reprod.* **37**:618–627.

Ziegler, T. E., Washabaugh, K. F., and Snowdon, C. T. 2004. Responsiveness of expectant male cotton-top tamarins, *Saguinus oedipus,* to mate's pregnancy. *Horm. Behav.* **45**:84–92.

Zinner, D., and Deschner, T. 2000. Sexual swellings in female *Hamadryas* baboons after male take-overs: "deceptive" swellings as a possible female counter-strategy against infanticide. *Am. J. Primatol.* **52**:157–168.

Zucker, E. L., and Clarke, M. R. 1998. Agonistic and affiliative relationships of adult female howlers (*Alouatta palliata*) in Costa Rica over a 4-year period. *Int. J. Primatol.* **19**:433–450.

Zucker, E. L., Clarke, M. R., and Glander, K. E. 2001. Body weights before and after first pregnancies of immigrant adult female mantled howling monkeys (*Alouatta palliata*) in Costa Rica. *Neotrop. Primates* **9**:57–60.

Zuckerman, S. 1932. *The Social Lives of Monkeys and Apes.* Routledge and Kegan Paul, London (reprinted in 1981).

Index

181